Yelping For Donald Trump

Avoid The Coming Collapse

Michael Mathiesen

Copyright 2016 - Michael Mathiesen

Dedication

All truths are easy to understand once they are discovered; the point is to discover them. - Galileo

There are only two mistakes one can make along the road to truth: not going all the way, and not starting. - Buddha

The truth is incontrovertible; malice may attack it, ignorance may deride it, but in the end, there it is. - Winston Churchill

In a time of universal deceit, telling the truth is a revolutionary act. - George Orwell

Truth is by nature self-evident. As soon as you remove the cobwebs of ignorance that surround it, it shines clear. - Gandhi

The great enemy of the truth is very often not the lie, deliberate, contrived and dishonest, but the myth, persistent, persuasive and unrealistic. - JFK

People can't change the truth, but the truth can change people. - Unknown

The Way of Truth is the Pathway to the Eternal Godhead. - Jesus Christ

You will know the truth, and the truth will make you free. - Jesus Christ

It takes two to speak truth -- one to speak, another to hear. - Henry David Thoreau

Table of Contents

Introduction -
Donald Trump is President

Chapter One   - What is Yelp?
Chapter Two   - What is Democracy?
Chapter Three   - How about something better?
Chapter Four   - From The Headlines
Chapter Five   - Yelpocracy In Action

INTRODUCTION

Donald Trump is now the President of the United States

What can you expect?

What about your investments? Are they safe? What about your home? Is it safe? What about your college education - will it matter? What about your job? Will you have one?

What about the country? Will there be one?

How will Trump Handle the Coming Total Economic Collapse?

FIRST OF ALL - Don't be too alarmed already. This is actually a book all about hope. Even if Donald Trump is Elected President of the United States - We The People have options to defeat the Brain Dead ideas that are going to flow from this man's brain.

I call it YELPOCRACY which is also the Title of an earlier book of mine.

Definition of Yelpocracy

Yelpocracy is a concept that originates from this author that is a combination of the Yelp 1 - 5 Star Rating System wherein users are able to rate their local restaurants, hotels, etc., and Democracy, the system by which people can choose their candidates and in some cases even proposed solutions to problems that originate from the people. Yelpocracy is the NEXT logical great change in society created by the modern technology of the Internet.

In the Yelpocracy Americans are now able to control their government and create common-sense solutions that become the law of the land. For example: Legalizing Marijuana on the Federal Level or Abolishing the IRS, according to the polls, easily merits a 5 Star Rating from the vast majority of Americans. Therefore, using this technology of the Internet in its highest and best use, we are no longer forced to live under the Dictatorship of unpopular ideas forced upon us by the power-elite.

Introduction

Since Donald Trump has declared Bankruptcy six times to avoid paying off his debts, we should assume that if Donald Trump is President of the United States, he will bring us to the brink of Bankruptcy faster than Hillary Clinton would.

It's going to happen during the next administration. The current economic system will collapse. Our economy will collapse soon because it's being artificially boosted by the Federal Reserve's printing money for decades. No nation in history has survived when their monetary system was abused like this, and no nation has ever abused its monetary system as much as this country has in history. We are currently about 20 Trillion Dollars in debt and this national debt that our government owes, and that means the debt that our government now has committed you and me to owe to the rest of the world is un-payable under any circumstances that are known to the human mind. The debt they imposed on us all, making us slaves to the banks, is growing at the rate of almost 4 billion dollars PER DAY!

This rate of growth is happening at a time when interest rates are at record lows of near ZERO. When the Fed raises interest rates on our debt, it will go up by nearly ONE TRILLION DOLLARS PER DAY. This rate of debt increase is obviously bogus enough so that everyone will finally get it. This, of course will trigger a total lack of confidence in the US Dollar, the currency that the entire world uses to pay their own debts and this means a total world-wide economic collapse is on the way.

There is no avoiding it. We've put off the day of reckoning by keeping interest rates at near Zero for years. Because they've artificially keep rates at near Zero it means that the pressure to raise interest rates is like a volcano ready to go off.

All of this debt of course stems from the over-spending of the last 50 years. We wanted to have a War in Viet Name, or should I say, a few of the Brain-Dead in Washington wanted that war. The vast majority of Americans did NOT want a war in a country on the other side of the planet, but we got one anyway.

*** - You will see me use the term "Brain-Dead" referring to the gangsters in Washington D.C. all throughout this book. This is not done out of a lack of respect for these people, it is out of TOTAL LACK of respect for these people, but also because it points up the fact that we in this country are controlled by a small group of people who have no ability or interest in doing anything for the average citizen. They propose solutions and pass laws ONLY to protect and serve those who give them the most money. This is why I call them the Brain-Dead. It's more accurate to call them traitors and criminals, but Brain-Dead just feels better and makes my point more directly. - ***

So, this book is primarily to also give the evidence that we are in a whole lot of trouble right now following the ideas of the Brain-Dead. Donald Trump is just another example of a small mind being in charge of us - when we could have a system of government that we all endorse and support and encourage and USE every day that would put much more people in front of us to elect to positions of power, but it can also be used to put many more and better ideas than those of the Brain-Dead in front of us for our total consensus and approval.

I call this a Yelpocracy where we rate all of the cock-a-mamie ideas that they put out there and only the best of the best finally merit our approval, the approval of the vast majority of us. This is a modern kind of Democracy wherein the ordinary citizens, you and me, would be served most of the time, instead of just the rich and famous.

We're all participating in the Yelpocracy even now, but we just have not connected up all the working parts to a form of Government - not until now. I have connected the pieces on the website that I created and it will serve as the birthplace of this new form of government where we are no longer forced to rely on the ideas of one man like Donald Trump - obviously part of the Brain-Dead Establishment - any longer.

Let him rant and rave for all eternity about building the wall, cutting taxes for corporations, keeping the IRS in power, etc. Let him praise Russia's President Vladimir Putin and even commit Treason by calling for Russia to continue hacking our country's most guarded secrets. Why not just JOIN RUSSIA as one of their slave-countries - Mr. Trump? I'm sure Putin would give you lots and lots of Real Estate deals under such a sell-out of our country.

Let Trump continue to insult women, immigrants, ALL minorities, young people, old people etc. Let him do all of these things because coming soon to a computer or smart phone near you is a way to CONTROL THE BRAIN DEAD and their BRAIN-DEAD IDEAS by offering up clear, concise, inexpensive, effective and beneficent ideas to the American people for them to RATE and then put the best of the best into law.

One thing is for sure, no matter what the politicians claim they are going to do for us - we can rely on the fact that they will simply NOT do what they say. This is made the truest of the true statements about politics because of the data presented to us by the last 50 years of this country's history. They promised they would keep us out of wars, during their campaigns to win their jobs of controlling us, and yet somehow we have had dozens of devastating wars and police actions all over the world that have cost us thousands of American lives and trillions of our dollars.

They promised us a great economy, lots of jobs, and yet, most of the good jobs have been exported to other country's all over the world who employ a slave labor type of system where the American worker, because we demand a decent standard of living, can not compete.

They promise us clean air to breathe and pure water to drink, but the air is no longer clean anywhere in this country and the water is not safe to drink, witness the billions of bottles of water that we are forced to buy today, walking past the city water provided to us at the plumbing tap.

They promise that they will improve and update our infra-structure, and yet we are forced to sit traffic every day for hours and if we go to a hospital with a serious disease, we are forced to either pay millions of dollars to the hospital or pay thousands of dollars to a Health Insurance provider.

During their campaigns, they all make promises they know they can't keep, and once elected they are ready with thousands of excuses why they couldn't keep their promises.

President Obama is the biggest example of this fraud on the American people. He campaigned on the notion that he would deliver "Real Change We Can Believe In". Millions of us bought the lie and he changed nothing - keeping all of the George W. Bush executive orders in place and reversing not one, even keeping all of our troops in Iraq for years before pulling them out, but even then pulling them out in such a way that he knew they would have to go back in there someday - or we would have even worse consequences. We are living with those consequences today.

Of course there is one thing that Donald Trump is right about 100% and it's the fact that "The Elections Are RIGED." Actually it's even worse than that. They rigged but also they don't count.

Our President's have always since the beginning of this nation been chosen by the Electoral College. These are people who are appointed by the states to actually choose who wins in each state. So, why we spend so much time and money watching the candidates debate, raise money and go through all of these twists and turns is beyond me. It's all just Kabuki Theatre a huge charade.

Now, I have to tell you here that I tricked you into buying this book. I also have a companion book entitled: "Yelping For Hillary Clinton." Hey - It's the same book, I just switched their names because this is the kind of Kabuki Theatre that I as a writer must perform to get enough Americans to buy this book before it's too late.

This is actually a book about the SOLUTION to the PROBLEM of always having to CHOOSE from the LESSER of TWO EVILS when in reality - in a true Democracy, we should be choosing from the BEST of the THOUSANDS upon THOUSANDS of US who want to be our leaders. And more than that, a way for us to create THOUSANDS MORE SOLUTIONS to PROBLEMS that thousands of American Citizens like you and me might have - ideas that are far more acceptable, less expensive, highly more effective and more responsive to the needs of the people, than the same old tired and corrupt ideas presented to us by the two gangs of political hoodlums - and then PICK the BEST OF THE BEST AGAIN that we can and must use.

Therefore, this is really a book about a concept I invented - Yelpocracy. I hope you will forgive me for tricking you, but it is the only way around the sustained suppression and oppression of my ideas by the major political parties and their henchmen in the media.

Life under Donald Trump as President would be just like all of the other Presidents we've had to endure over the years. He will come up with totally un-acceptable ideas to the vast majority of the American people, and then somehow Congress will endorse some - reject others, depending on their impact on the major moneyed class in this country. NO IDEAS of Trumps will be put into effect until and unless they don't hurt any of their Pay-MASTERS AND ONLY IF THEY AID THEIR Pay-MASTERS.

We all know this. We see it day after day in this country and we are unable to do anything about it because the system we live under gives us no path to correct these egregious crimes against humanity.

UNTIL TODAY -

One glaring example taken from the HEADLINES TODAY requires your consideration here - In a later chapter I have dozens of examples of how they screw us over and use their power only to SERVE THEIR Pay-MASTERS.

The Federal Government continues to put our citizens in prison - CONFISCATE ALL OF THEIR ASSEETS - for smoking a weed known as Marijuana. Even though Marijuana has been researched thoroughly over the years to the point that 22 states make Marijuana available to their citizens to be used as a MEDICINE that is proven to cure many of the world's most deadly diseases, the Federal Government - because they are paid by the major Pharmaceutical companies, the biggest drug dealers in the world - the President and the Congress continue to claim that Marijuana is harmful and many millions of people in this country are still duped enough by this effort to continue to believe this fraud.
MAKE IT CLEAR - MANY IN THE FEDERAL GOVERNMENT HAVE BEEN CAUGHT SMOKING POT - but since they are PAID to SUPPRESS IT because it would seriously impact the PROFITS of their Pay-Masters, they continue to PUT MILLIONS OF AMERICAN IN PRISON FOR SMOKING IT - even though they smoke it in private.

THIS IS THE best example I can give about why a NEW and IMPROVED TYPE OF DEMOCRACY is so essential at this time in our history - BEFORE WE COLLAPSE on our own malevolence by our leaders. There are hundreds of others you will learn about in this book.

Hundreds of polls have found that over 90% of the American people want Marijuana to be taken OFF the CONTROLLED SUBSTANCES LIST so that people could smoke it or ingest it as a medicine anywhere in the country without fear of committing a crime in these United States - the so-called bastion of Democracy.

SO, it's NOT LIKE THEY DON'T KNOW THE WILL OF THE PEOPLE EITHER.

*** There is hope because you purchased this book. ***

This is an urgent appeal for you to help kickstart the Yelpocracy I describe in great detail wherein the people of this country - and later the world - can take control of their country and later their planet for the progress, prosperity, freedom and justice of ALL MANKIND - not just a lucky few who can and will control us all in this modern form of SLAVERY that we must all endure.

Even as we write these words, President Obama is on TV making a speech and he just said, "We in America are all about Democracy." And "You've got to want Democracy." The only problem with these statements is that he himself doesn't believe in Democracy or else he would have given it to us by now.

AND in fact, Obama did try in the first few months in office in his first term to bring a Yelpocracy to the forefront of his administration. On WhiteHouse.Gov, he placed a Yelpocracy where ordinary citizens could make proposals that the government would consider and then the rest of us would RATE THESE POPULAR PROPOSALS.
BUT WHEN THE LEGALIZATION OF MARIJUANA rose to the TOP of the White House Suggestion Box - the entire site was taken down. There were no further comments about this little experiment in Yelpocracy. The media made no note of its passing. It just ceased to be something that Obama would support any longer.

Today, you can create Petitions for other Americans to sign at WhiteHouse.gov. However, if you try to POST a REAL HELPFUL IDEA that might impact the PROFITS of any MAJOR CORPORATION - YOUR PETITION WILL NOT APPEAR.

They - those who control us - don't really want Democracy in this country but - trust me - but it's coming anyway. It's going to arrive in the form of a Yelpocracy and just keep improving to something we all realize delivers finally the full potential of a truly democratic society.

AND YOU ARE GOING TO HELP. Let's all start Yelping about it. Go to http://www.yelpocracy.com Start Rating this book and encourage everyone you know to do it at our Yelp Rating Page.

https://www.yelp.com/biz/yelpocracy-santa-cruz

ALSO - if you can find a Donald Trump or Hillary Clinton Web site where they have OPTED-IN to any kind of online rating system - GO THERE and DO THAT.

* * *

When most of us are asked what type of government our country is defined as - we would reply that the United States of America is a 'Democracy'. Yet, we are not. We were formed as a Republic, the same type of government that was created by the Roman Empire in order to give the Roman people the impression that they were in charge, whereas in reality they were totally under the control of the Emperor.

The form of government we are forced to endure today is the creation of less than 100 men who were farmers and owned hundreds of slaves to work their farms. They could not create a United States Constitution that included real Democracy because the first thing that the people would have done would have been to abolish slavery. We know this because the vast majority of people who lived in Colonial America were not white farmers and slave-owners. The majority were workers in dark dusty factories and small farms. They were abused economically. They had extremely poor working conditions. They were forced to work under extremely dangerous and toxic conditions even from childhood. There is no doubt that most Americans would not support slavery if they enjoyed a Democracy as early as 1789.

The United States Constitution therefore is based upon the principles that only a wealthy elite should rule over us because their own wealth will always be threatened by the majority of Americans who would, in almost all cases, would make decisions that was more equitable and fair to their own interests, the interests of the common man, the worker class.

When slavery was finally abolished by the 13th Amendment to this original slavery-protecting Constitution, the African-American slaves were freed, however the basic fabric of our society still enables slavery and the types of slaves who were easily identifiable by the color of their skin have merely been replaced by all of us taking their place. We work daily in jobs we are forced to endure and pay part of our incomes to the government without any voice in the matter. We are forced to travel to our places of slave labor in vehicles that are designed to keep us separated and suspicious of each other, without any voice in the matter. We are forced to breathe polluted air, drink polluted water, even to think in polluted ideas and again with no voice in the matter.

In Congress, the so-called 'Seat of Democracy' our elected officials spew forth ideas that may sound necessary and important for the health and welfare of us all, but the best ideas are never even put to a vote in Congress. One or two dozen of these same people are paid by Lobbyists to reject solutions that would do the most of us the greatest amount of benefit because the greatest benefits to society would lower the profits of the multi-national corporations, the evil employers of the lobbyist community.

We no longer even have to wonder about the type of system we live under when we look at how Obamacare was forced upon the American people. This 1,000 page of legislation would be put into effect without the Congress even reading it. And, under this same law forced upon the country by President Obama, the IRS is given the power to enforce it. This means that an American citizen can currently go to jail for choosing not to purchase a health insurance product from one of the three or four corporations who offer it. In a real democracy, force of this nature can not and must never be used to make a citizen do anything - unless the law in question is endorsed by the vast majority of citizens. When only one man shapes the laws of the land, and uses the government to enforce his sole ideas that the rest of us must live under - this is clearly not democracy - it is Dictatorship.

Would there have been many more choices for solutions to the health care problem if the system were opened up to the involvement of the American people? I believe that if the question of solving the health care crisis were put to the American people, thousands of much better ideas would have been proposed by the American people themselves. IF we then enjoyed a Yelpocracy type of RATING OF THESE POPULAR IDEAS - the best of them would have risen to the top of the heap and then we would have had much better choices - and I'm sure that none of these proposed solutions from ordinary people would include the enforcement by the IRS. No sane citizen would want to give the IRS more power than they have already over our lives.

When only one person or a small group of people create a law that they know is unpopular only these types of people would require such a strict and foreboding body like this. It's ironic that the first American Revolution was begun under the slogan of 'No taxation without Representation'.

Today, we have one thousand times more taxation, than in Colonial America, and there is today absolutely no voice in the matter from those who are taxed. When is the last time you got to vote on whether or not we should have a tax on our incomes?

Of course, both Dictator's gangs are complicit in the conspiracy to deny the American people their true freedoms. When George W. Bush took office in his first few months of his administration, he alone decided that it was a bad idea for the small business person to be allowed to send email to people they were soliciting to buy their products and services. The only reason that a man who never had an original idea in his head, had this one was because he was taking bribes by the large corporations to think this way because if the small business person could email everyone, the large corporations would be forced to compete with thousands of folks instead of just a few of their friends who constituted their monopolistic price-fixing clubs.

So, George W. Bush, with help of the same bribed Congress called this new form of freedom of expression and freedom to compete 'Spam' and made it illegal. Since that day, the American economy began to sputter and slow down to the brink of total collapse that we're suffering today. This one man has caused millions of Americans to lose their homes in the worst economy since the Great Depression, an economic condition that he alone accomplished with help of the United States Congress who are also bribed by the same corporations that got to Bush. Millions more of us who have not lost our homes are still struggling and working two or three jobs to make ends meet and just keep our families together. As I write these words, 45% of the American people are on some form of Prescribed pain killers, mainly because life has become so harsh in America. We've completely lost the optimism and the ways of the American Dream because of this one man. Under his predecessor, Bill Clinton, the economy was in near perfect condition with full employment and wages rising faster than a rocket.

Of course this fraud on the American people was just a minor warm-up misdemeanor compared to the felony that

President George W. Bush committed against the American people and the Iraqi people. After the attacks on the World Trade Center, Bush needed a scapegoat. He couldn't tell us the truth, that the Saudi Arabian people were responsible for the attack on the United States since 18 of the 19 people who hijacked the airplanes on 9/11 were Saudi Arabian citizens of King Faruk.

Bush couldn't tell us the truth and direct our anger against Saudi Arabia because the Bush families has billions of dollars invested in many Saudi Companies and he knew that if he attacked that country Bush's assets would be in danger of dropping in value by those billions. So, he concocted the story that Saddam Hussein the President of Iraq was close to getting or already had nuclear bombs that he could use against America. In truth, Iraq had no such technology. They barely had enough industrial expertise to keep their oil pumping. Bush actually told a lie and got thousands of US troops killed in a war that was completely fraudulent, cost us trillions of taxpayer dollars and took the focus off of his assets in Saudi Arabia

In a real Democracy, this most heinous crimes eve3r committed by a political leader, except the ones by the homicidal maniac and psychopath Adolph Hitler would NEVER have been allowed to happen in the first place, and if it did, the perpetrator would be in prison by now.

This was of course, just another warm up for the next heinous crime imposed upon us by President Bush. When the banks failed during the Bush second administration, instead of prosecuting the executives who committed the crimes of cheating millions of account holders all over the world - Bush rewarded them with Bail-Outs and Bonuses. This was the equivalent of saying to the biggest bank robbers in history - hey, we know you needed the money so we'll give you a couple more billion dollars that we will steal from the American people.

How can one man gain so much evil control over the rest of us. Today, we get just as much unsolicited email as ever before, but now it comes from foreign entrepreneurs who live and work outside the jurisdiction of American laws.

Bush in one stroke of the pen, made it illegal for Americans to make a living as independent small business people and made it legal for foreign entities to take over this role - the greatest force in the American economy is now destroyed by this one man. How can that same man later be allowed to lie to us all about where the true attackers of this country came from? How can one man use his power and influence to reward those who do us the most harm and are the nation's biggest criminals?

One man can do so much harm to us all because we have NO WAY OF STOPPING THESE CROOKS. Until now.

In a democracy, this type of over-whelming and obviously bad idea would have been put to the people for their consideration and final adjudication. In the Yelpocracy, we would have been shown thousands of better ways to control the problem of unwanted emails than the complete eradication of a segment of society that the American Economy had embraced and endorsed for centuries.

How would you rate this proposal over the one George W. Bush gave us that made it a criminal act to send someone an email?

Proposal: To solve the unwanted email problem, We The People declare that anyone found sending an email that includes the possibility of a fraud being endorsed or supported in that email, the sender shall be subject to a one million dollar fine and no less than ten years in prison. If a foreign entity is found to be the source of a fraudulent email, that entity shall be banned from using the Internet.

Give a 1 to 5 Star Rating to the above please.

Compared to the one that George W. Bush forced on us, I'm confident that a much less damaging proposal to the incomes of millions of American small Businesses - the engine of this economy - would have risen to the top of the heap of ideas and eventually prevailed.

One more glaring example of how and why we are all suffering a dictatorship and a total control by the money-elite.

Marijuana is now approved and legal to prescribe as a medicine in 22 states of these United States. This means that this herb has been researched by the state health care systems in these same states and found to be healthy and completely effective in treating many of the most dangerous diseases in our society.
At the same time, polls have shown that more than 90% OF THE AMERICAN PEOPLE FAVOR THE LEGALIZATION OF MARIJUANA.
Yet, somehow the Federal Government has decided that they know better than the health departments of 22 states, and all the other states are soon to follow, that they know best. They continue to use billions of our dollars to follow Americans around, spy on us, arrest us and put us in prison for years for doing something that is completely healthy and beneficial. Why do they take this extremely unpopular stand?
They are bribed by the American pharmaceutical corporations the biggest drug dealers in the world to keep it illegal because these same corporations have noticed a steep decline in sales of their drugs that are far less effective than Marijuana and sometimes 1,000 times more expensive. The legalization of Marijuana when it becomes Federal Law will cost these legalized drug pushers hundreds of billions of dollars. Therefore, in their minds, it's cheap to invest a few million dollars to bribe our elected officials in Congress to keep as much pressure on this very benign substance as long as they can, so that they personally can take billions from us in the form of BONUSES - being paid for my the American Taxpayer.

How would YOU rate the notion that Marijuana should become legal and no longer a criminal offense?

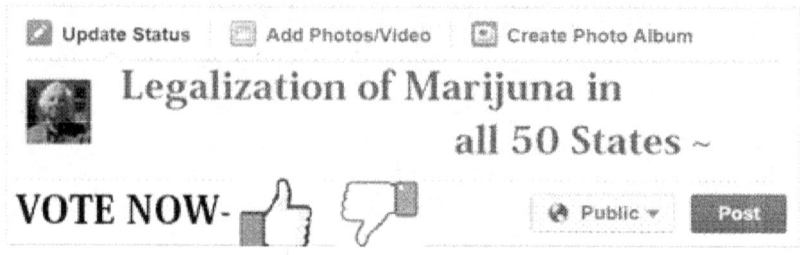

Please Give a 1 - 5 Star Rating Please

Hopefully you are already wondering why we are not already a Yelpocracy. In a later chapter we give you dozens of great examples of how we can and MUST reverse some of this Dictatorships dumbest ideas such as The NSA Spying on us, another great idea from George W. Bush and maintained under the man who said he'd change everything and Foreign Aid now almost ONE TRILLION DOLLARS PER YEAR of your money and your children's money being given away freely to foreign nations - even those who we are at war with - a great idea?

You may have even better ways in mind to solve these or other problems that we face today as Americans.

Yelpocracy is the only way to get the best ideas to the forefront of our awareness where we can easily sort through them using the modern algorithm technology we use every day to find the best restaurants and hotels as judged by our peers, and not by the force of advertising dollars.

If you can appreciate the simplicity and the beauty of putting the most modern and effecting technology in the history of this world to the purpose of controlling the laws and policies that the world must endure, you must follow your appreciation to its logical conclusion and demand that we begin such a system.

In rating this book - you take Step One in the process of delivering True Freedom, Liberty and Justice for all to the American people. You become a Yelpocrat and you become a founding partner in the only sustainable future of this country and the world outside of our borders.

Upon completing this book - Please visit our website and make your rating count. http://www.yelpocracy.com

The world will thank you.

Chapter One
- What Is Yelp?

When you go out to eat at a local restaurant, chances are you or someone in your party has looked up the Yelp Rating for the restaurant in question. This is accomplished by searching the World Wide Web for the web site of the restaurant you will be visiting that day or evening and looking at the combined total of all the 1 - 5 Star Ratings that customers have attributed to that restaurant over the years.

It may look like this on your computer or smart phone:

Shadowbrook Restaurant Capitola - Capitola, CA | OpenTable
www.opentable.com › ... › Capitola ▼ OpenTable ▼
★★★★⯪ Rating: 4.6 - 4,714 reviews - Price range: $30 and under
Book now at **Shadowbrook Restaurant** Capitola in Capitola, explore menu, see photos and read 4716 reviews: "The entire experience far exceeded all ...

This is an average of all 5,000 reviews given to one of my favorite restaurants - Shadowbrook - almost a perfect 5 stars.

This almost a perfect Five-Star rating means that like everything in life, they're not perfect and one or two out of a thousand customers are not going to enjoy their visit there as much as anyone else. They may have experienced an over-worked or dispirited server that evening and so they were not as impressed as they would have liked.

You will almost never see a perfect Five-Star Rating on any restaurant because everyone is going to have an off night once in a while.

Hopefully, in the case of your own restaurant or other business - you will strive for Five-Star Ratings from the vast majority of your customers. In this way, you can be assured that your restaurant will always be appreciated as you intended for a long time into the future.

Here's an example of what the Yelpocracy would look like for YOU and ME.

Mike Mathiesen - Yelpocracy to Save Our Country - & The World Later
www.Yelpocracy.com   America 2.0 Inc
★★★★½ Rating: 4.6 - 4,714 reviews -
End the TYRANNY of the Two-Party System. Inventor of New Rating System for new SYSTEM for selecting our TOP LEADERS and BEST Political Solutions.

Not only would we rate the Political Candidates in this way, but we would also RATE THE RATERS. Why not allocate more VOTE POWER to those who have proven themselves to be more active in the system than someone who is too young and inexperienced to be able to choose the best solutions compared to an older person like me who has written several books on the subject and spent a large portion of his life active in the revolution itself?

To me, this is a key element of the Yelpocracy. We need to give more political power to people who have a greater Democracy Quotient than someone who has none.

In most elections in this country, the majority percentage is a mere 51% of the voting electorate to pass or reject an idea. That means that most of the time, the country is divided right down the middle. There will be almost HALF of the country FOR something and another HALF AGAINST.

BUT, what if the majority to pass something was set at a higher standard - say 60% or even 70% of the voting public? That's probably going to work out much better for the system in the long run as people get used to the idea that if they are in the minority - at least something was accomplished and they may even be encouraged to change their mind about that particular solution they voted against, as they see it working effectively over time.

And, we are not talking about just using a 5-Star Rating System such as Yelp exclusively to make our choices for the Yelpocracy. We will use the best features of Yelp, Twitter, Facebook, Linkedin, Reddit, Quora and others because they each have qualities that are helpful in sorting out the bad from the good.

AND more importantly my Yelpocracy Rating is going to be weighted by how good a RATER I am with other people and ideas. In other words, my Yelpocracy rating is NOT just what other people think about me and how they rate me, but also a comparative scale as measured against other people in my category and how well I have rated others in relation to how they appear to others.

For Example: IF I get a notion to go out and BASH SOMEONE because I don't care for the color of their skin or something they said, then I am likely to give that person a LOW - ONE STAR Yelpocracy rating. However, 99.99% of everyone else gives this same person a 5 Star Rating. Therefore, my rating is an OUTLYER, an anomaly, outside of what everyone else is rating this person or place. Therefore, I am not objective in this case. AND, this OUTLYER rating of mine, where I let my ANGER, DISTRUST, PARANOIA, PERSONAL PREJUDICE or whatever rule my judgment - will LOWER MY OWN OVER-ALL Yelpocracy Rating. In other words, I shot myself in the foot by UNDER-RATING someone who to all the rest of the world is a Five-Star Personality. Over time, I'm going to learn this lesson in Yelpocracy and have a much more restrained point of view in making any more RATINGS of others.

"DO UNTO OTHERS as you would have them DO UNTO YOU." As a very HIGHLY RATED person once said over two thousand years ago. It applies just as much today in this modern world as it did back then, and maybe even more so.

SO, I as the RATER am also being rated by the ill-advised and undemocratic things that I do in haste, indifference, or antipathy for my fellow man.

We have to always remember this in the world of Yelpocracy or else we are only going to replace one bad system with another.

Too many restaurants have gone out of business, jobs lost, people's good reputations ruined by a band of online hoodlums who decide to trash a place simply because one of their gang had a beef with that restaurant. Anyone can have a bad night, even the best restaurant in town has them. So, it's not fair that you get bashed by a group who think it's perfectly moral for them to bash a place out of existence because their ego was bruised by a surly wait staff in one night, when millions of happy customers never have this experience.

Yet, this happens many times using the present form of Yelping. HOWEVER, if we rate the rater, then we are going to see that this ONE BASHING was not justified because 99% of customers never have the same problem - then this BASHING is 1. Not recorded for the public to see. And 2. Backfires and lowers the rating of the HOODLUM who gave the unjustified rating.

If this concept of Rating the Rater were applied to all of those POOR RESTAURANTS who were put out of business by malicious bands of online thugs, most of them would have survived this modern day form of vandalism.

DO NOT FORGET - RATING THE RATER - IS one of the KEY and FUNDAMENTAL FEATURES of the Yelpocracy or else we can have the same kind of influencers, people with indiscriminate followers, gangs, political parties, lobbyists, special interest groups, mobsters, big corporations, etc. running things just as they do now.

So, Yelpocracy is very familiar to us all. It's very close to what we do every day online. We give a like or a thumbs-up to our friends when they post a picture or a thought that we admire and want others to see. When we give something our like - we also expect that others will follow our lead and like the same things. Many times they do and sometimes they do not.

In the case of a major UPGRADE to YELP, the Yelpocracy, we would simply add the concept that the person RATING the other person must be held accountable for those times when we're being bad, hasty, naughty, annoyed, frustrated, angry or just plain mad enough at some body or some thing that we are forced to take it out on someone totally innocent. These fits and rages, that we've all had should not destroy another innocent person's livelihood and sometimes the many hundreds of other families who may be dependent on that restaurant being around.

In the case of a new political system, we have to take special care that there are no over-rateds getting into office, such as the George W. Bush's and the Barack Obama's. Here are two perfect example of people highly over-rated by their respective parties, and neither of whom were fit to lead the Free World. Yet, this is the kind of person we get every four years and we must always choose between the lesser of two evils, instead of choosing from the best of many thousands of applicants.

This Presidential Year is an excellent case in point.

Hillary Clinton in conjunction with her husband Bill Clinton has used her position as Secretary of State to sell positions in the State Department and probably other branches of government to the highest bidders, even foreign bidders. This has been proven by the publication of her private emails. There may be other conflicts of interest that We The People are unaware of. More news is coming out about this as we write this blog.

Donald Trump - on the other hand - has shown himself from his own statements to be sold out to Russian business interests, who will even go to the trouble of hacking Hillary Clinton's emails with Trump's support so that he and PUTIN'S RUSSIA CAN BUILD A HUGE PERSONAL Empire for themselves and their cronies, family and friends even larger than the one Bill and Hillary are building.

The evidence of this comes in the recent news that Paul Manafort, Donald Trump's former campaign manager took millions of dollars from the Russians and channeled it to Lobbyists here in this country.

The only reason to give money to Lobbyists is to have the Lobby for those who give them the money - in this case, the Russians. This is probably why Trump has been saying very flattering and favorable things about the International thug and Russian Dictator - Vladimir Putin.

This is probably the worst choice between two evils we've ever had to suffer in the history of this country. Either way - whichever candidate wins the White House - the American People are SCREWED.

There has to be a better way to choose our leaders and I believe this BETTER WAY is through the use of Internet Technology and I am going to use the method that YELP uses for our purposes of helping everyone understand how this better way COULD WORK - however - the EVENTUAL MACHINERY may have nothing to do with YELP in the END. We may have to invent our own system - however - as a DEMONSTRATION of the BETTER WAY to choose our candidates - YELP provides the best example.

We use YELP to HELP us choose the BEST RESTAURANTS, Auto repair, Hotels and other service businesses that we want to utilize - WHY NOT USE the highest and best technology of the INTERNET to choose who will lead this country and others out of these Dark Ages and into a much better Future for ALL Mankind?

Please BOOKMARK this page and SUBMIT IT TO EVERY SOCIAL MEDIA ACCOUNT YOU participate in - so that WHEN LAUNCHED

The Yelpocracy may have a chance to supersede the antiquated and corrupt system we must all suffer today.

Yelpocracy will definitely bring about the clamor of thousands of people who want to apply for the job of 'Top Dog' or the leader of the Free World. Why not give thousands of people the chance to bring us their ideas. We must never trounce any of them without giving them all an equal playing field. If we rate the raters of new ideas, then we can weight every opinion in a manner that will total up to the most objective and accurate rating system that has ever been devised by human kind.

There will be flaws and people will find ways to subvert them, just because there are always those who will want to go back to the ways of the horse and buggy for no other reason than the greed for more money and power, something probably deeply ingrained unfortunately in our DNA.

We must remain vigilant that whenever uncovered, we fix the subversion as fast as possible.

Where does this idea come from?

The Highest State of Consciousness yet devised - The FeedBack Loop

To that END - While we are busy CONSTRUCTING the YELPOCRACY

Please read more about how to give the proper FEEDBACK for this or any idea that may spring forth at this time in our Evolution.

From This Author

The 4 States Of Consciousness

4StatesOfConsciousness.com

How Did Everything Get So Entangled

AND HOW to UN-Tangle Them

Chapter Two
- What is Democracy

Before we can really appreciate the why's and wherefore's of a real Yelpocracy in place and performing the heavy lifting of all or at least most of our governmental decisions, it makes sense that we should take a closer look at the history and the evolution of government because when studied objectively, it is a fascinating tale indeed, and one we need to be fully briefed on as Citizen/Yelpocrats. T
he first government, during the time when we were all mostly cave dwellers, was most probably a Patriarchy. A tribal leader, usually the strongest cave man would take up his club and declare himself to be the leader and much like a pack of wolves, a few other young cubs might occasionally challenge this self-proclaimed leadership of the biggest and the strongest. And so probably this type of tribal leader was overthrown every few years as younger and stronger bucks came up against the leader and overthrew him in a bloody contest with sticks and clubs.
    Messy to say the least, but effective and brief revolutions like this made for a kind of order that the rest of the tribe could easily follow and obey. Then, after we became more agrarian and stopped chasing our food every day, around 10,000 years ago, with the development of proper housing or huts for the people who had now become farmers, roads would develop in between the most important huts and then even groups of these important huts where people who specialized in butchery, bakery and candle-stick making would be concentrated thus forming the earliest communities.
    A system of governance for these early farmers and merchants would develop into what was later known as City-States where a tribal leader would bestow upon himself the Divine Right of Kings and Queens, wherein, they would rule with an iron hand with the aid of hired goon squads, we now know as Armies, Navies, Air Forces and Marines. Basically, this military power, kept the young bucks from overthrowing the tribal ruler for decades and sometimes even

for generations, which lead to the dynastic powers that we know as the Egyptian Empire, the Greek Empire, the Persian Empire, the Chinese Empire, the Roman Empire etc. An empire being simply a King who was bold enough to put into the laws of the land that his next of kin should become the next King or Queen upon his or her death.

The rule of Kings and Queens even extends to modern times where there are still Kings ruling over Saudi Arabia, Brunei, Great Britain (this power mostly symbolic) and indeed, my research tells me that there are a total of 31 countries out of about 277 nations on the Earth even today ruled by so-called Royalty.

Still more than 10% of the world's nations being ruled by folks who believe they are somehow stronger, better, wiser, more eligible to rule as the entire government in their region and their descendants to inherit this right for all eternity, no matter how good or bad they end up as a ruler. In a kingdom there is no accountability, no recourse by the people, no redress of grievances, no freedom of the press, freedom of speech or any other freedom we hold so dear in this country today and in many other countries.

The fact that these people rule with an iron fist over their countrymen does not escape most of us. If it were not for the military power that supports them, all of these people, or at least the vast majority of them, would be deposed almost instantly upon removal of the iron fist, their goon squad, armies, navies, secret police, etc.

And, in fact, if it were not for the American Revolution of 1776, probably many more positions of the world's populations would still be ruled in this heavy-handed way where the average citizen has no power to object to the rules of the road, no power to a fair trial when accused of a crime, no control even over his own person which is technically owned by the Sovereign Lord. Slavery is another word for it. We didn't care for that much in the British Colonies of three centuries ago and so we took to the streets, took on the British goon squads of the time, known as Redcoats, and as every school child all over the world now knows, won our freedom and independence.

This is how we got to where we are. Some of the bravest and brightest bulbs in the pack got the notion that we should all be equal under the law, that slavery was not a good thing and that there was no real Divine Right of Kings as the Kings had proclaimed for thousands of years.

Indeed, our founding fathers stated quite clearly and emphatically that if God was to get involved in human affairs, He would surely be on the side of the ordinary citizen, the family man or woman who was given life and that this life would belong to God and no one else. So, the Divine Right of Kings was out the window in 1776 for most of us and replaced by the right to be government being bestowed upon elected officials in what had come to be known as early as Ancient Greece as a "Democracy", a rule Of, By and For the People themselves and therefore a right that they themselves having bestowed could take back or alter in any way they might choose and at any time, without having to clear with anyone other than themselves.

The only flaw in the original form of Democracy in ancient Greece is that only a few proposals submitted by the elders of the City-States would make it to a vote of the majority of citizens. There was no way for an ordinary worker to propose a law. They all came from a central group of leaders who hung out together in a small building that eventually became the Acropolis and they would discuss in the 'Socratic' method, which proposed changes to the rule of law they would give to the people. There was good enough reason to perform this top-down approach to Democracy at the time due to the fact, that the vast majority of people alive at this time never had any kind of education and did not read or write. Only a small elite group would be taught how to read and write and from this, there arose a small cadre of philosophers who would come up with all of the ideas that would be put forth for the consideration of their people.

Another major flaw of early Democracy as it was invented, was the fact that a good sized minority, close to half of the Greek population were slaves, people who had been captured during time of war, brought back to Greece where they would serve the upper Crust of society as waiters, butlers, construction laborers etc. Obviously, these people were not allowed to vote.

Be this as it may, this flawed invention and the authority that Democracy gave to ordinary people is the basis of our government at least in America and a few other imitators in Europe and elsewhere and although flawed in many respects, it is still the best thing, in my humble opinion, that anyone has yet to create in the business of ruling over each other, creating the laws of the land and enforcing them and even giving a tribe of people their unique identity in the world as the fighters and defenders of freedom even unto the rest of the world. A great and noble cause, America is then, is it not?

And aren't we amongst the proudest of tribes to call ourselves Americans today? And isn't our system of government a proven best by the fact that millions of other oppressed people try to sneak in here every day? Is there any reason to question the continued existence of such a great and noble cause? I think not. However, it is not done.

Our democracy is only a half-way house in the quest for real democracy, isn't it? The founding fathers who created our democracy were afraid to some extent of the granting of too much power to the people too fast, so accustomed they were to living under the rule of kings. It seemed almost unnatural that ordinary people would have as much power as they themselves, since they, our founding fathers owned large tracts of lands, some of them as big as states are today. And, nearly all of them owned slaves. Thomas Jefferson owned about 100 slaves at the time of his input into the United States Constitution and James Madison owned about 300 slaves.

John Adams, George Washington, James Madison, Alexander Hamilton, the signers of our basic framework of government all owned slaves, but eventually gave them their freedom, after realizing this dichotomy and the cognitive dissonance of proclaiming freedom to all white men, they had forgotten for a short while that even black men deserved freedom too under their own logic. Having nowhere else to go, most of these early freed slaves stayed on to work for their masters, but this time being given a wage, a few pennies per day and a roof over their heads and free medical care, such as it was.

However, our own basic framework, the United States Constitution, the basic law we live under today, made slavery itself legal because they knew that their friends who owned large estates and farms relied on those slaves for their own livelihood and they simply could not imagine a better way to get the crops to market in an efficient yet humanistic manner.

Therefore, the basic framework of our government was and still is based upon the concept that you cannot have a real democracy because the slave class in that society will immediately begin to vote themselves more and more rights and a larger and larger share in the way that things are done. Slavery of the poor unfortunate people whom we abducted in the middle of the night under threat of violence by the slave trade was not outlawed until the 13th Amendment in 1865 and passed with the aid of President Abraham Lincoln, still today one of our greatest Presidents, because he saw to the completion of the list of freedoms that was guaranteed to us in a free and democratic society.

NOW THIS IS A MAJOR POINT I AM MAKING HERE.

Slavery still exists in this country today, however, because the basic mechanism of making our decisions still lies in the hands of the few to protect them from the power of the majority of people. One class of slaves has merely been replaced by a new class of slaves. Slaves walk around our country today, but instead of being dark in skin tone, slaves sport every kind of skin tone there is and every kind of educational level as well. The so-called 'Middle-Class' in America has become the modern slave class. How can I say this in all candor and authority?
The answer is simple. Just ask yourself if you agree with anything your government has done in the last 50 years. I guarantee that the vast majority of you will say no. Therein, is the proof of our slavery conditions. They never ask us our opinion on anything, except after they have made the decisions for us and then they give us fine speeches about why we should have supported their wisest decisions and these exculpatory speeches usually starts out with 'Well, my fellow Americans, we did the best we could, but we had to compromise with the other side.'.
There is no other side. We are the other side, and yet as far as I can remember, the ordinary American people have never been asked to compromise on anything they decided to do to us, or things would only be half as bad as they are.
Under our slave-Constitution, the one we live under presently, the one where there is NO MENTION of the PEOPLE VOTING DIRECTLY on the ISSUES and thus bypassing and/or over-ruling some of the most egregious and heinous decisions ever made in

history by this so-called REPRESENTATIVE form of a 'Republic' we will always be slaves.

OUR DEVELOPMENT HAS BEEN ARRESTED

The evolution of Government really has stopped there - at the point of a REPUBLICAN form of Democracy. And at least we can thank God that our founding fathers also gave us the right and the methodology to UPGRADE our flawed system of government any time we so choose.

This is such a time.

Since 1865, our Constitution really hasn't been improved other than to give women the right to vote in the 1920's. BUT NEITHER WHITE PERSON OR BLACK, MALES or FEMALES, YOUNG OR OLD have gained any more rights or basic freedoms truly since that time, more than a century ago. We still have an unfinished democracy because also created by our Constitution besides legalized slavery was the manner in which we elect our representatives in Congress and how they represent us based on population, the House of Representatives, and apportioned by states, in the United States Senate.

The average citizen's interests, it was stated and still pertains today would be safeguarded by these elected United States Representatives and Senators and the will of the people, or at least the vast majority of us expressed by these same people and put down and codified into the laws of the land by these same people given their authority to do so by US, We the People.

For a couple hundred years, this system has served us well enough. However, recent events suggest to many of us that we are no longer served by these same elected representatives nor the system that made them thus, giving them a new kind of Divine Right of Power. Kings and Queens it seems have been replaced by pompous and arrogant white men in business suits who are elected to office after months spoon-feeding us quaint little sound bites to help us think of them as good folks just like us. But, as soon s they are in office they change their spots and start listening to the lobbyists more than the people who elected them to office.

Even descendants of these same rulers have claimed their rights to these same offices. I love the Kennedy family, but who would argue that they are not a dynasty of power and rule over us? Al Gore, the son of a Senator almost made it to President of the Untied States and indeed is a good person. Largely because of his work for the environment, I like Al Gore very much, but who would argue that someone else outside his family would have done as good a job? George W. Bush, the son of a Director of the CIA who went on to become President felt that his claim to the White House was a Divine Right and who would argue that anyone else could have done a better job from 2000 to 2008?

Getting elected to office has itself come to be a big business. Campaigns for these same offices now run into the HUNDREDS OF MILLIONS OF DOLLARS, which few of us can afford. So, people have emerged from the shadows who know where large pots of money are and they beg the support of these people and entities and spend it all on so much advertising and marketing that most of us get fooled every four years into thinking that this time, we found someone who will represent us.

Indeed, each and every candidate uses that vaunted and clichéd phrase to help convince us it is true, when the truth is that they will leave these offices as multi-millionaires, sometimes even billionaires and to the great detriment of our laws, policies and rules that we are forced to endure in our daily lives. If you want proof of this total lack of representation, you only have to read the laws that they have enacted over the last fifty years. Almost none of them reflect the true interests of the average citizen, no matter what state or region they are from.

Everyone should agree that Congress, for example has the mission of regulating Wall Street, but as Bernie Sanders (I) from Vermont said the other day, "Nowadays, it's the other way around. Wall Street regulates Congress."

There are a few notable exceptions. Social Security is one of this country's greatest achievements, but that set of laws were passed during World War II under Franklin D. Roosevelt, a man who really was a man of the people and probably the last one of that type. Even though from a wealthy family, he felt that the little guy was not getting an even break and said so many times in many ways and most of our modern laws that truly represent the needs of the people have their origins in this one great man.

Most Presidents and most of Congress have busied themselves reversing these great advancements in civilization. George W. AMBUSH even being so bold as to start the privatization of our great Retirement plans and making them subject to the whims of the stock market. Later machinations by Bush would teach us why he wanted to do this. The transfer of our payroll taxes into Wall Street Brokerage firms where they would have managed our accounts for us would have made them TRILLIONS OF DOLLARS in fees and commissions and their track record such as it is, probably to no great advantage to our retirement incomes at all and more likely to the great trimming of our benefits.

And, as far as the White House goes, there is no mention in the Constitution that the President shall have the right to make war on any foreign nation. Yet, Lyndon Johnson did it. Richard Nixon did it. George W. AmBush did it and so did his father. So, truly what we have as far as National In-Security is concerned is a ONE-MAN RULE. Lyndon Johnson and Richard Nixon did this to us in Viet Nam as well.

This is not exactly what I would call democracy in action and so in a way, we have evolved full circle back to the Divine Right of Kings and Queens when our founding fathers gave us the rule of the ordinary citizen over ourselves, known as a "Democracy". The fact that they labeled the first version of America as a Republic matters not because the Republic for which we all pledge our allegiance as school children was based on the principles of Democratic rule and the promise of more advanced features of democracy have been written in stone and in blood into all of our actions in amending the Constitution up to this day.

So, our Constitution is a living and evolving document set up to be a living and evolving document wherein We The People have every right and even a duty to change and adapt at our will. If this were not true, slavery would still be legal and acceptable today, women would still be second class citizens and the ordinary worker in this country would have no rights, we would have no middle class and we would all be working in some way, shape or fashion for our lords and masters.

Instead, we have had the ability over the years to extend our rights and privileges and we have done so under this limited form of democracy bringing us ever closer and closer to the dream of a full expression of democracy and democratic principles. Where you give freedom a chance to grow, wherever you plant the seed of freedom and equality and love of country, this seed will take root and it will grow into a great tree of life.

The founding fathers understood this great natural principle and they planted that seed, but the growth of the tree is under attack from the pesticides of modern forms of the Divine Right of Kings and Queens who want to rule over us and rule over us forever. Something has to give. Or else we are all doomed to repeat history over and over where the oppressed eventually get angry enough to overthrow their oppressors and it's usually not pretty and it can lead to even worse state of affairs.

We must be careful and ever vigilant to not make this great mistake of history and trust those who advocate the violent overthrow of our oppressors because in this case, at this point in history, our oppressors are us. We have done this to ourselves by the simple act of omission. Through our apathy, the forces of evil have taken their power over us. By simply turning from apathy to activism and choice, we can move this great country back into the proper tradition of more freedom, more democracy, more optimism and success.

It seems to me, however, that these choices that we make in the near future must include a basic and fundamental change to the system itself. Hiring another one of these cronies to do our bidding is not enough. We have suffered such a long string of disappointments that it is now clear that We the People must take a closer and stronger hold of the reins.

The only way to do that is to make our own political and economic system more responsive and more agile in bringing about the most benefit to the most people most of the time. This is our only hope. The greatest debates in politics for the last century or so has been the great debate about Socialism vs. Capitalism. There are plus sides to each system of government of course. The positive side of Capitalism is that it allows the individual to succeed or fail in life based on his or her own merits and work activities. The negative side of Capitalism is that it provides for a Boom or Bust type economy.

And boy are we in one of the worst BUSTS in history at the moment. We are now growing our national debt at the rate of hundreds of millions of dollars per hour. Federal spending is out of control. They are simply printing checks now to finance every ridiculous project they can think of - even importing more people into this country whose homes and homelands, our military - at an earlier out of control expense - destroyed.

Sadly it could have all been avoided too, millions of people might not have lost their homes, thousands of unnecessary deaths could have been prevented except for the lack of leadership and morality from those who truly should be in charge. We are now known as the 'Silent Majority' strictly because we have no way to voice our opinions or to take control.

In the last century or so, we've had to major Booms and two major Busts, the last one having started during the Bush Administration for reasons we explain in a later chapter on the excesses of Power. But, this Boom and Bust cycle of Capitalism has another negative side effect in that these cycles produce a greater and greater gap between the richest strata of society and the poorest.

Today, for example, wealthiest amongst us are able to afford dozens of homes all over the world, private jets, a caviar and champagne lifestyle with little worries about their future. The poorest segment of our society, by far the largest segment, and this rift wouldn't be so bad if it were reversed with many times more people in the wealthy category of society than in the poorest strata, but alas it is not reversed.

The lowest strata of our society is combined from vast masses of us who are struggling to feed our children and make the payments on our houses or apartments and the fear of being jobless and homeless is currently rampant all across the land because the vast majority of us don't get the chances to score big in business, the stock market, real estate, etc. and we run out of time.

We get older and slower, or we get sick and can't work as hard as we once could. So, the negative side of pure Capitalism is that people are treated like cattle and are abused by the Capitalists who take advantage of their poor bargaining skills and make profits from the daily labors of the poorest of us without providing us with any real safety nets or any real future or even hope for the future.

Of course, we don't enjoy a pure Capitalistic state any more. Today, we are somewhere in between pure Socialism and pure Capitalism because we provide many safety nets for our poorest classes. However, there are now more safety nets for this lower segments that the upper segments can little afford to keep it going in the black, so we have been forced to borrow from the future generations of rich and poor to support the massive social programs of today. And to such an extent now that we are inevitably headed for bankruptcy, if we're not already there. Being propped up artificially by China, a combination itself of Communism and Capitalism, and other countries, this is not a good thing for our future freedoms and traditions.

The plus side of Socialism is that there are such safety nets in place that everyone is guaranteed something to eat and a place to stay along with completely free medical care. The power of the economic decision-making to sustain such a huge cost on society is place in the hands of a small bureaucracy, who have little incentive to get things right, being that they can never be removed from office. There are no freedoms of speech or of the press, so if anyone wants to make a complaint or even offer up a suggestion, this citizen can find himself imprisoned or shot in the head without a trial, as we have seen in China currently and in the Soviet Union in the recent past.

The People being the greatest force of change no matter how much you oppress them, this complete system of Socialism has either collapsed all over the world wherever it has been applied such as in the Soviet Union, East Germany, Poland, Eastern Europe, soon Cuba, etc. or it has adapted and changed to more modern versions of Socialism such as in China where the allow the people to make the basic decisions about where to work and so forth, but they also limit free speech and there is no freedom of the press or judicial freedom and you can still be shot in the head for objecting to any of this.

However, it should be noted, and it certainly has been noted by billions in China and around the world that it is the engine of Freedom of Choice and Capitalism that has made China the greatest economic power in the world, recently surpassing that role of the United States. The Chinese economy is exploding at around 10% per year, something unheard of in the West, solely because the Chinese Communist Party has loosened their Maoist grip on society and let them work for their own personal successes.

This most basic need to succeed and provide some of the modern goods and luxuries of today's society is what spurs most of us on to create and work hard in society, providing jobs for other people and thus stimulating the revenues of the state and making their country rich in terms of money, economic growth and power, and especially in personal achievement.

These two simple economic factors always seem to go hand in hand in creating the best economies the world has ever known. The key then, it seems to me is to continue this trend toward combining the best of both political and economic models.

In an earlier book of mine, I make the argument that someday we should even consider the Incorporation of the United States of America.

Why? Because this one simple action would instantly make all of our leaders - ACCOUNTABLE - LITERALLY. They would have to balance the books. They would have to have quarterly reports to the shareholders. They would have to work under a much higher standard of transparency than we hold them to now.

And, above all else, it would make every single citizen a 'SHARE-HOLDER' in the over-all productivity of the nation. We could create entire streams of income based on our relative impact on the bottom line of our own economy. We would each be more productive and happy in our daily chores knowing that we would someday participate on a fair basis compared to how much we put into the system over our lifetimes.

I'm hopeful that Yelpocracy may someday even evolve to this highest form of government - a 'For-Profit Corporation', because when you think about it, - what is the only way in the long term to prevent the leaders of any nation from stealing the wealth of nations? And that one solution - it seems to me is to treat them more like hired hands instead of knighted demi-gods, Aristocracy and major celebrities.

If we incorporate our government, we use the best part of Capitalism, the Corporate model of growth, change, job creation, profit making, business ethics, governance, etc. Our leaders would have to provide a real resume to be considered for the most powerful office in the world. We would no longer pick them from random chance and/or the filtering process of a few people known as the Party Professionals who always ending up choosing one of them. And in a Yelpocracy, everyone would be chosen by the millions of us who use the Internet every day to choose the best restaurants, hotels and other business services, trusting to the real-life experiences of those who have been there, done that, gotten the T-shirts. No longer would we have to accept their FALSE and MISLEADING ADVERTISING CLAIMS.

And, by incorporating, we also have the chance to allow everyone to share in the wealth, the equity that is created by society acting as a whole and pulling in the same direction that Socialism promises, without any of the negative sides of each system used in and of itself. So, the best of both systems are struggling now to maintain some kind of equilibrium. But, the next step eludes us all. Currently our leaders in government and business are squabbling amongst themselves and in public over every little detail.

The current debate is raging around how to afford Health Care for everyone in this country and how to regulate the banks, reverse the massive Bonuses that the financial corps are paying to all the cronies, the same people who acted so unethically and so recklessly to cause the pain and suffering now felt all over the world and so that the Second Great Depression doesn't continue for much longer or to prevent a third. These squabbles over minor petty details about how to run a nation, how to rebuild the economy, how to share in the equity of society is a squabble that obscures the greater issue: How to bring about real changes, real changes that everyone can appreciate and enjoy and that will last into the far off future of Humanity?

The evolution of business into the modern day corporation and the evolution of the modern day form of government can and must now merge and unite to bring us one clean and simple over arching system of governance that could not be 'Gamed' as the modern politicians have 'gamed' the system that exists today. In this one simple and graceful motion to a more common sense approach to managing the way we do business in this country and the way we control our society, we accomplish the instant hybridization of both of the greatest economic philosophies in the history of Mankind.

In this way, we combine the best of both worlds and rid ourselves of the negative sides of both as well, maximizing the best and most positive sides of both systems. We can have it all. We can have nearly everyone become wealthy and prosperous and healthy and rested and relieved of the stresses of life. And, we can have a booming economy, employing our greatest talents, utilizing all of our best ideas, most creative works, hard work and due diligence nearly one hundred percent of the time because everyone would be a shareholder in the total economic output.

We still have the right to fail because if we all goof off and take too many vacations or forget how the system is designed or try to abuse it, like anything else created by humans, it will also fail, and we will not participate in the profits and we will become poorer for it, but it will be ALL OF US WHO FAIL TOGETHER, from TOP TO BOTTOM, instead of just the poorest of us failing to thrive in these circumstances and that is the basic difference that we can have if we want it. Tremendous success for everyone, if we want it. Or a shared experience of losses if we do not work together and in harmony.

I leave it to future generations to decide which way they want it to go? But until we achieve a system where everyone shares equally in the successes and failures of our society as a whole, there can be no Real Democracy, there can be no really favorable ultimate destiny of Mankind, because if we do not improve our system of making these kinds of decisions we will simply continue to destroy our planet, piece by piece, as we are now doing under the present systems of uncontrolled, mismanaged and unregulated human greed, and then what? Where do we go from there?

In the next chapter, I give you what I am calling the Beta Test version of Yelpocracy. I will show you how we can combine the evolution of this flow of better and better ideas through all of our history and the evolution of the Internet so that they intersect at a place that can not only save our country from all threats both foreign and domestic, but which would also save our species from the total annihilation and extinction we all face from the combined activities of our own selves.

Remember - Yelpocracy strives to utilize at all times the best of both worlds, the political awareness we have inherited through the centuries and the technological savvy and creativeness that we have squeezed out of the last few decades.

Chapter Three

- The Beta Test

Chapter Four

- From the Headlines

Hopefully, I have at the very least, whetted your appetite for more nutrition in this healthy diet of new ideas. So, in this chapter, I want to give you hard and cold examples of how we could have used the Yelpocracy in the last several years, and how we MUST BEGIN to use it in the next few years or else - we're doomed to continue doing the same old insane things to ourselves every day until we fail completely as a nation and maybe even a species.

Many scholars call the United States of America an ongoing experiment in democracy, because that is exactly what history teaches us. It is not a completed form of governance, and there never will be such a thing especially in this new era where technology changes the very essence of our relationships to one another almost daily. And so this Doomsday Experiment will build upon the earlier model experiment, but it does so with an increased sense of urgency and the awareness that we cannot afford the snails pace of change that we've been forced to suffer since the inception of the first experiment in democracy.

Today, we are charged with the ability for everyone now reading these words to use this same experimental process, mankind's greatest talent, to achieve even greater things and especially in the area of extending the American experiment in Democracy. It will never be perfect, but in the ongoing learning process, we will always get closer and closer to perfection, as we fine tune the process and build mechanisms to prevent total disaster.

* * *

    This is the practical part of this experiment and forms the basis of our vows - that we will FIGHT the EMPIRE with the most powerful weapon of all TRUTH and Truth only. Herein lies HOW we can achieve so much and the reason that I have spelled it all out for you as the most solemn part of our vows is that we now have all of these tools that we can use to get ourselves out of the worst pickle we've ever known in all of our 5 million year climb up out of the jungle and into cities and towns. By this point, you should be all fired up and ready to go. You are starting to feel the collective consciousness or if not - you soon will begin to feel it. Everyone catches on in their own schedule and in their own way. I know you are catching on and this is my greatest accomplishment to date and one for which I am the most grateful and humbled.

    But, you need more direction. You may even have the Cosmic Connection in place already that I am trying to instill in every one of my volunteers. But, I also know that some do not have that yet and need just a little bit more encouragement - a few more details on how we can accomplish so much that we would be saving our planet and our own species from the ultimate and conclusive disaster.

    Well, the answer to the biggest question in your mind is that we will use the power that we have just unleashed in 'You' and by that I mean the collective 'You' or all of you, the communicating and very powerful union of those of you who have volunteered to see this to the end. This is the 'Collective Consciousness' that we are testing here and it is now up to you to be part of it or to ignore it. Being part of it is now made easier by the invention of the Internet as I have said, but in this chapter I am going to give you how I can see the Collective Consciousness working in much greater detail and it is at this close-up magnification of this Knighthood that I hope to capture your participation once and for all time.

    By placing our full faith and trust in the greatest tool for democracy to ever land here on planet Earth, we can and must start to turn our world from its present course to total annihilation and destruction of all life forms on this planet towards a final destination of total peace, harmony and global sustainable economy and husbanding of our resources in such a way that we guarantee all future generations of humans and all other life forms an always-improving environment upon which to live.

Please remember that this line of reasoning and the accompanying chores that we ask of you is based on the following question:

Why in the name of all that's holy - do we allow massive decisions effecting the entire planet - the fate of over 7 billion people and countless billions of plants and animals upon which we depend be made under the mental capacity of only a few men (even fewer women) in suits? Why do we rely on this antiquated method of governing ourselves when the tool for managing all of this for the benefit of all 7 billion of us is right before our eyes?

You will want the proof.

In a previous book - "Using Google Super-Vote to Create The Super States of America" I detail how an Internet democracy could work with almost no major alterations to the Internet and how we use it today. So, I won't take the time to detail the nuts and bolts of a Real Democracy using the Internet except to say that if we can use Google Super-Vote where American TV viewers vote on their favorite singing contestant on American Idol and other reality TV shows, then, it's obvious this same technology could be used to vote for our most important proposed solutions to our most pressing problems, not only for who should win a million dollars.

Another consideration that people have whenever I mention this is that we do not have a Democracy in America because it was technically set up as a Republic. Although this is true historically, the same history books prove to us that when people want a major change in their system of government, they merely band together and demand it. Eventually it gets done if they are persistent and have the courage of their convictions. Mohatma Gandhi, the man who single-handedly brought a form of democracy to India and freed his country from British tyranny said "First, they ignore you. Then, they laugh at you. Then, they fight you. Then, you win."

Many scholars call the United States of America an ongoing experiment in democracy, because that is exactly what history teaches us. It is not a completed form of governance, and there never will be such a thing especially in this new era where technology changes the very essence of our relationships to one another almost daily. And so our Knighthood will build upon the earlier model experiment, but it does so with an increased sense of urgency and the awareness that we cannot afford the snails pace of change that we've been forced to suffer since the inception of the first experiment in democracy.

Today, we are charged with the ability for everyone now reading these words to use this same experimental process, mankind's greatest talent, to achieve even greater things and especially in the area of extending the American experiment in Democracy. It will never be perfect, but in the ongoing learning process, we will always get closer and closer to perfection, as we fine tune the process and build mechanisms to prevent total disaster.

* * *

The following are examples of how we can and MUST BEGIN to use

the Internet in the world's first YELPOCRACY to untangle and solve our nation's and later the world's most perplexing problems.

And as YELPOCRATS - I am hopeful that 'YOU' will think of many others for this is our most pressing battle in the war against the Evil Empire of confusion, ignorance and GREED. There will be others.

The only way to do that peacefully is to PUT all of the MAJOR CHOICES we must make as a NATION on the INTERNET for ordinary people to resolve for ourselves, thus rendering the power of the Empire as less and less until it is finally defeated by common sense solutions and not ones that they were PAID TO FORCE UPON US by the special interests.

With our FINAL VERSION OF YELPOCRACY -- or FILTERING the best candidates and IDEAS and then Voting On The Internet we could instantly have the following laws approved by the voters within minutes and at ZERO expense. All of this is known fact by the use of the latest polling. Therefore, I am giving you the best laws that would be produced by this highest and best use of the Internet, because we already know how the majority of American voters feel about these issues.

AND my personal favorite ideas are listed here. These are taken from the recent history of our country and my ideas on how to solve related problems in a far more effective and less expensive way than our government structure attempts to squeeze solutions to our problems out of an out-dated and obsolete horse and buggy system.

In this basic format, you would be VOTING on the IDEAS themselves - HOWEVER - in the BETA-TEST of the new and improved problem solving system of government we would be merely GIVING our RATINGS to these ideas and helping to push them to the top of our communal awareness.

SO - HERE WE GO - NUMBER ONE because in recent POLLS 90% of the American People wanted to STOP CHASING pot-smokers and incarcerating them for doing far less than having a beer.

YET, our Federal Government continues to classify Marijuana in the same classification as HEROIN - a Class-One Controlled substance even though 22 states in this country classify it as a MEDICINE with proven HEALTH BENEFITS.

1. Marijuana would be taken off the list of criminal substances and even legalized everywhere just as alcohol was a century ago. This is supported by 90% of the American people, yet Congress wants to keep it a crime. Why? Because they are bought and paid for by the money-class in this case - the MAJOR DRUG PUSHERS of the WORLD - the PRESCRIPTION DRUG COMPANIES. In this issue they are bribed to make no changes to CRIMINAL LAWS by the Pharmaceutical companies who know that over half of their revenues would be gone due to the healthier benefits of a decriminalized marijuana ingredient, Cannibinoids.

AND THIS FORM OF DEMOCRACY IS EASY TO UNDERSTAND when you see it in this familiar way - something we already DO on a daily basis for TRIVIAL IDEAS. Real Democracy is SIMPLE TODAY and this proposal would look like this -

VOTE NOW-

SEE HOW SIMPLE REAL DEMOCRACY CAN BE?

Because the government has been locking people up for decades for using this stimulant while at the same time allowing millions of Americans to die horrible deaths from Alcohol related diseases and auto accidents, we know that they're lying to us when they say it's a dangerous drug, when there is not one death that can be attributed to cannabis. Pot actually makes people very peace-loving and calm and serene and without fear. You never hear about a bar fight amongst pot-heads. The government can only exist in its present form if they make us all fear them. Therefore, the real reason they despise this drug is because it makes too many of us no longer fear them and that spells the real danger for those in power over us.

Additionally, we spend billions of dollars per year housing people in prisons for imbibing in this harmless product. And, most of the reason, this is allowed to go on is because the Pharmaceutical companies are our real law-makers, along with the oil companies who keep us using their products, the banks who force us to go deeper and deeper into debt every day and those in our government who have been bought and paid for by these special interests and many others. Even foreign governments have their spies in Washington D.C., who daily contribute to the law-makers bank accounts. Can our law-makers represent us any more when they take money from foreign interests? This is why The power of the Internet must be employed to give them their direction and when they disobey the will of the people, they must be terminated.

2. We would surely vote to cease all foreign aid until such time as we could afford it again. This proposal again is supported by at least 90% of the American voters, yet Congress does nothing to curtail foreign aid - a form of welfare to foreign rulers, and in fact is increasing the budget outflow of our taxpayer dollars, and borrowed dollars, every year to other nations, most of which hate us. This makes no sense in terms of a real democracy where the vast majority of us are completely ignored. Show me an example where our foreign Welfare Assistance has done us any good and I may change my mind, but you cannot.

This money gets passed by Congress as aid to the people of these foreign nations, but when it arrives it always gets re-routed to the personal Swiss bank accounts of the gang-leader who happens to be in power at the time - never reaching the real need and just enriching criminals.

3. We could vote to raise taxes on corporations who do business in this country. We could do this in numerous ways even raising taxes on them and allocating the FUNDS to SOMETHING that COUNTER-ACTS the ALL HARM they and their INDUSTRY DOES to the rest of us.

Currently, corporations avoid the income tax rules by holding meetings in other countries and therefore claiming they are not based in the United States, therefore, BILLIONS of taxes are lost to this fraud. Our law-makers could easily reverse this unfair and ridiculous slap in the face of American tax laws by simply passing a law that requires that every dollar earned in the USA is taxable at 10%, no deductions, no loopholes. Calculate it every year and send in the checks. Currently, Apple Computer has declared the largest profits in the history of the world. Yet, they pay no taxes on this income because they now claim to be headquartered in Ireland. And, they create very few jobs in this country because all of their products are made by Chinese girls in such deplorable conditions where hundreds are committing suicide every month. They pay them next to nothing. They also hoard their profits and don't give much back to their investors in the greediest hoarding of capital ever known, thus doing the American economy almost no benefit at all, only taking from its consumers, more than triple what it could charge for the same products.

The only reason our law-makers have not done this years ago is because they are bought and paid for by the corporations who they regulate. They are also given high-paying jobs when they retire from government by the same corporations who also bribed them during their terms in office. This is the biggest pay-offs they can give our law-makers because they all end up with 5 or even 10 million dollars in salaries from the Pharmaceutical companies, the Oil companies, the Big Banks and they are not even required to come to work every day. A meeting or two over the phone once or twice a year is all they need to do to "earn" this money and it lasts for the rest of their lives. Big pay-offs like this are common because in not passing laws like the decriminalization of Marijuana, for example, they allow the pharmaceutical corporations to continue making BILLIONS.

The same goes for the oil companies. By not regulating them so that they pay their fair share of taxes into the US Treasury, they are allowed to make billions. Paying off a few key Senators or Congress people a few million is nothing in comparison and in our system, which these corporations love to death, they only have to pay off a Committee Chairman because he or she can prevent a proposed law from even coming to a vote.

Which brings me to Number Four.

4. Using the Internet, we can make it law that "No committee in Congress shall have the power to block any legislation from coming to a vote in Congress. Any Senator or Congress person may introduce a bill in Congress and Congress must then vote on it, up or down, within 14 days of its introduction on the floor. A speech of a bills introduction made by any member of Congress shall be sufficient to introduce it formally to Congress and this speech is sufficient to start the clock."

You realize that CONGRESS will NEVER pass any legislation that takes power away from THEM and this is why we need to FIGHT THIS BATTLE against this form of Corruption.

Congress will never ABOLISH the IRS either - no matter how many Presidential Candidates PROMISE they will do this - because it's all of the present LOOPHOLES that enables them to TAKE MONEY from those who USE the LOOPHOLES. When you go to a FLAT TAX - there are no LOOPHOLES and therefore, NO MONEY will be PAID to Congress to KEEP THEM WIDE OPEN and this is another example of WHY And HOW we must and CAN FIGHT THEM.

One more provision of this law states.

"Any member of Congress shall be granted sufficient speaking time for proposed bill introductions every month so as to carry out the intent of this law."

5. The abolition of the personal income tax.

NOW - you may see the LIGHT at the end of the TUNNEL - and HOW and WHY we must use the INTERNET if you have not seen it YET.

This is one of the most heinous of all tyrannies in this country today. Taxes on our incomes if they were administered fairly and objectively would be fine, but the millions of pages of code that has been developed over the years, mainly due to special interests, makes the personal income tax a burden on all citizens who are forced to go through complex calculations every year to be in compliance.

Instead the Income tax rules could be abolished by one simple sentence - and replaced with a second sentence that set up a National Sales Tax of say 10% of anything sold in this country. The IRS could remain in business but enforcing this new tax code that replaces the old one.

Here again, all the latest polls confirm that at least 90% of the American people would gladly vote PRO on such a proposal. And again, the only reason we have not changed our primary method of taxation from over-complex personal income taxation to a flat taxation of consumption is because special interests are allowed to bribe our law-making body.

The National Association of Realtors, for example, pays millions to Congress people to keep the Personal Income Tax rules in place because they hold the very narrow and selfish view that if we didn't have the personal income tax DEDUCTION for HOME MORTGAGE INTEREST that fewer people would buy Real Estate.

This narrowly focused belief by a wealthy class of people who number in the smallest of minorities, Realtors, prevents them from seeing that if there were no income taxes, a much vaster population would have sufficient money to buy Real Estate without any further inducements.

Another major example of how we need to drastically curtail and even eliminate the inequities and the iniquity in our law-making process is one of my favorite pass times and that is imagining how much we would have frustrated the evil that one man can do to the world, the biggest example of this being the administration of one George W. Bush, who in his infinite wisdom or lack of same make all of the following decisions almost totally single-handedly. Even Adolph Hitler would have been impressed at this American President's accomplishments in the destruction of real democracy in America.

First days in office Bush declares that we are not allowed to email people whom we don't know. This was the Can Spam Act and Bush jammed it through Congress in the first month or two of his administration. Obviously, he was bought off by the large corporations who were not able to compete with people working from their home office, as I was doing at the time. So, they made it a criminal act to try and follow the American Dream of Success. This was the start of the fall of our economy. Because at the same time Bush made it illegal to call anyone on the phone whom you did not know. These two laws made it impossible for the small entrepreneur to get a leg up in the business world and become a success with very little money invested. The economy has never recovered and never will until we ---

6. Repeal the Can Spam Act and the Do Not Call list so that anyone can begin a business from their home once again without fear of going to jail and paying huge fines. Since this insane and criminal act was forced on us by George W. Bush, we know its origin is part of the Evil Empire and must be reversed. This is one of the many things I thought Obama would repeal in his first few days in office - but he has so far, left all of George W. Bushes policies intact.

THEY LIE TO US EVERY TIME THEY SPEAK

SO HOW EASY IT WOULD BE TO REVERSE any EMPEROR'S PRONOUNCEMENTS?

ONLY the INTERNET allows us this much POWER RIGHT NOW - AT NO EXTRA COST to the country.

Of course, the use of the email system and phone system to commit a fraud on any American citizen shall remain in effect. But, if you have a legitimate product or service, in the Free Enterprise system that made America the greatest country in the world, this process of finding a market must be unencumbered in this way.

If someone is annoyed at receiving my emails, he or she can simply delete it, use a spam filter, or file a complaint to the government that can still maintain its power to regulate a fraudulent individual or company. But as long as I have a product or service that people want to use every day, I must be allowed to present it to the world without fear of my government trying to preclude my ethical, moral and legitimate right to my own economic means of support.

George W. Bush went on to commit much greater evil. When he started to talk about going to War in Iraq due to Saddam Hussein possessing Weapons of Mass Destruction, there should have been an Internet Vote on whether or not we needed more verification of these 'suspicions' by the Bush Administration.

A waiting period or cooling off period before an American President makes the decision to send troops would be a wise law to have in effect given the two major wars in my lifetime - namely Viet Nam and Iraq - that American Presidents have caused and that have proven to be unnecessary and expensive in terms of both American lives and the American treasury now depleted because of the George W. Bush Fraud upon the world, the biggest fraud in history.

A period say - thirty days before any troops could be committed by the President - during which time, a thorough and independent investigation by the American people is performed to tell us if in fact with physical proof that Saddam Hussein possessed these 'Weapons of Mass Destruction' as Bush loved to state, would have saved our country tens of thousands of our own war casualties, and more than two trillion dollars.

On top of these direct costs of the Bush Fraud, the most expensive costs are yet to be visited on the American people since Bush also failed at his war and the succeeding President Obama also failed and left a power vacuum into which all of the craziest of crazy terrorist groups have flowed into and to such a degree that has allowed them to create their own state which we will now have to fight for decades, creating even more unnecessary casualties and even more trillions in treasure. However, because our Treasury is now conveniently depleted by the even bigger Bush Fraud that followed the WMD Fraud - the Bank Fraud and subsequent bail-outs, we no longer have the power to go and eradicate this menace that we might have had, if we had not acted so recklessly simply because we had no way of preventing George W. Bush to carry out the full completion of this fraud and we certainly had no way to prevent it happening over and over again.

This kind of complete and total reckless disregard for the safety and security of the American people must be completely surgically removed, this cancer completely and totally obliterated from the American system of governance and decision making.

This can and must be done now in this generation of living Americans or the entire world may be destroyed.

7. Another great example of how we reorganize our country is to repudiate or refuse to pay the horrific National Debt that they have thrown down around us, strangling the US Economy. The national debt totals almost twenty trillion dollars as of the writing of this book and is skyrocketing by another 500 Million Dollars every few hours. It can never be paid off, unless they start to tax every American man, woman and child more than 100% of their incomes. Of course, few of us would sustain such a thing and so it is impossible in any realistic scenario where this debt gets paid off. It can only continue to sky-rocket and continue its drag on the economy.

THE ONLY WAY TO GET OURSELVES OUT OF THIS MESS is to PENALIZE THE WEALTHY CLASS WHO GOT US INTO IT IN THE FIRST PLACE -

This proposal will be labeled as "COMMUNISM" by those who would be the most effected. HOWEVER, unless we want to declare the largest BANKRUPTCY in HISTORY and throw the world into the greatest DEPRESSION OF ALL TIME - this is the only HONORABLE WAY TO PAY OFF THE DEBT that THE WEALTHIEST AMONGST US CREATED BY INVESTING IN IT.

TO KEEP this unsupportable DEBT on the BOOKS for future GENERATIONS TO PAY OFF IS SLAVERY OF THE UNBORN. A real democracy doesn't stand for that in the least.

AND remember this DEBT was CAUSED MOSTLY BY WALL STREET - PRESIDENT'S BUSH and OBAMA who believed FOOLISHLY that WAR is the ANSWER to our ECONOMIC PROBLEMS which is the OPPOSITE of the TRUTH and which has cost us TRILLIONS with NOTHING GAINED for it.

THIS proposal would tax the ASSETS of all wealthy individuals and corporations to such a DEGREE that it would cause a hardship to them, but also pose the LEASE HARDSHIP to the rest of us, the vast majority in this country who can barely afford to stay in their own homes - which are EXEMPT from this LEVY.

PAINFUL YES - and HIGHLY UNDESIRABLE but made necessary BY THE RULING CLASS the only supporting underpinnings of the EVIL EMPIRE.

Because of this huge burden we are currently unable to repair bridges, schools, hospitals, roads, what is known as the infra-structure of this country, the backbone of any real society. We are also unable to defend ourselves. When the Commander-In-Chief should be sending troops to quell the murderous terrorists in Iraq and Syria, he can only send a few ineffective jet planes who drop bombs on abandoned buildings so that they can show this activity to the American people who are duped into thinking we're winning this war when in fact we're losing. We can't even afford to counter the Russian advances in the Ukraine and Russia is in worse economic condition than we are. We can't even defend ourselves against children who wander across our borders daily. Imagine how vulnerable we would be if attacked by a real army.

Therefore, a law must be proposed on the Internet and therefore, in due time, the Federal Election Ballot where we all vote on a reasonable payment plan for the American people Since, we are at most, ten percent responsible for this debt, in that we have blindly gone down this path with our law-makers and done nothing to stop them, we should be responsible enough to pay ten percent of it.

This, I believe would be supported again, by at least 90% of the American electorate because, it's simply the only way out of this economic crisis they've put us in - always on the verge of economic disaster. It's amazing to me that we are able to sustain this untenable position for so long now. Ever since Viet Nam, We The People have been forced to finance the insane projects of tyrannical Presidents supported by the co-conspiracy of the Congress. That stops now.

In giving the National Debt this severe hair-cut, no sane investor would ever loan us this amount of money ever again, or at least until such a time that they realized that with the American people finally in control of their financial well-being, there is an even greater chance of getting repaid than there is currently and we would have the full faith and credit of the world once again, but we would never be forced to borrow money for the insane pet projects of Presidents any more such as Viet Nam, Iraq, Afghanistan and unlimited entitlements.

Today, more than half of the American people are receiving government assistance in some form. This is clearly unsustainable, is causing most of our current economic trouble as we go more and more into debt to pay for these programs, and it makes us all wards of the state as in a Communist country and this - as we have seen - doesn't work in the long run.

8. IN ORDER TO PREVENT THIS KIND OF ECONOMIC CALAMITY EVER AGAIN another National Proposal that would arise over the Internet as soon as we begin the process would be a proposed law in which Congress is not paid their salaries or any other benefits in any year in which the Federal Budget is not in perfect balance. In other words, you want to support ideas that cause us to go into debt, we don't pay you your salaries because above all, we don't want to pay for your insanity. As soon as one of their paychecks has been suspended, they'll find a way to balance the budget, hopefully by dropping all of their insane practices such as paying college professors millions of dollars to study the sex lives of snails.

Don't laugh this is one of the recent laws they put into effect. Don't get me wrong, there's absolutely nothing wrong with a biology professor wanting to learn more about the sex lives of snails. However, since most of the sex lives of snails is now known, let him delve into further prurient behaviors such as this on his own nickel. You and I certainly should not be required to pay for all of this nonsense that totals in the billions of dollars every year.

9. We might use the power of The Force to even change the monetary system towards a more equitable system. It's obvious to some that the Federal Income tax was instituted by the large banks who were and still are the most powerful of people in this country. J.P. Morgan, Rockefeller and other robber barons, in 1913, met in secrecy and began the Federal Reserve System. At the same time, they bribed our law-makers to pass the Federal Income Tax legislation that we are all slaves to today. The obvious connection to the big banks are that if they take some of our money from every pay-check, they can use that cash flow as NATIONAL RESERVES upon which they can issue debt instruments, United States Bonds and Treasury bills, that are now peddled all around the world. The more of these debt instruments are created, the richer the banks become because the interest on the debt is paid to them. Of course, there has to be a better way to create money, one that would keep our nation out of debt and put us on a firm economic foundation that would employ more of us and enrich our daily lives in a more equitable manner.

It's no surprise to me that the top one percent of our population earns more than all the other 99% of us combined. This is due to the banker's ownership of our economy. Everything we do as a National Economic system is geared towards their benefit more than to any other group. This is not sustainable in a democratic society and will be changed using The Force, if I have anything to say about it.

10. Another example from the pages of history. Santanaya said "Those who forget about history are doomed to repeat it." What statement holds more truth than this? After the Viet Nam fiasco, I thought, well, maybe this country learned something from that ILLEGAL AND IMMORAL INVASION OF ANOTHER COUNTRY. War is the last resort and you cannot use the power of the American military to invade a small country without resources because you just can't. This is what the world is trying to stop. But did we learn? Oh no. George W. Bush invaded Iraq under the exact came FRAUD of lying to the American people about Weapons of Mass Destruction, the same exact manner in which Johnson and then Nixon lied about the threat to America by the one of the tiniest countries in the world, some fifty years prior. Our government learned nothing from Viet Nam and now hundreds of thousands of our youngest and finest have paid the price. We're all doomed unless we start to remember history all the time.

IF AT THE TIME that BUSH was PROPOSING such a NIGHTMARE SCENARIO as we see today - and GIVEN a Real Democracy in America - WE THE PEOPLE could have VETOED THIS LUNATIC'S IDEA.

RESULTS - TWO TRILLION to TWENTY TRILLION DOLLARS SAVED - because OBAMA wouldn't still be there defending Bush's war.

AND millions of LIVES SAVED.

The same thing could have been done during the Lyndon Johnson and Tricky Dick Nixon's similar EVILS DROPPED on VIET NAM.

And, you can find more at my original thought on this subject - "America 2.0 Inc. - Take stock In America" found on Amazon and on Audible.com

The Following law should be put up for a vote by the American people over and over until such time as it passes the will of the majority. History has taught us too many times that Presidents can, if allowed to go unchecked, will abuse their powers. Therefore, it should be possible for the American people to impeach any member of Congress and the President of the United States, and perhaps even a Supreme Court Justice with a two-thirds majority vote of the electorate. If we were given this power in the Viet Nam era, we would have saved millions of lives and millions of square miles of pristine rain forests, now gone forever due to the bombing of them with Agent Orange, a toxic chemical still leaching into the oceans. All of this done under the orders of one psycho-path - first Lyndon Baynes Johnson and then to an even more harmful degree by Tricky Dick Nixon.

If we had the power to impeach the insane people who make it to positions of power over us, there would be far fewer examples of this massive destruction of our planet, our animals and our innocents. George W. Bush, emboldened by the fact that none of his predecessors other than Nixon were impeached for their crimes, took the level of Presidential crime to its highest power by invading another country based on lies, and allowing our true enemy - Osama Bin Laden - to go free - an act of Treason, and conspired with his cronies on Wall Street to defraud all the banks around the world. A tall charge, but why else would he then pay them billions of dollars of taxpayer money as their reward for committing such crimes rather than prosecute them?

If politicians knew that We The People were watching their every move and had the power to not only impeach them, but make them pay for their crimes, there would be far fewer of them that we are forced to pay for.

Think about it. Why should the American People be constantly forced to pay for their criminal behavior. In an honest and robust form of Justice, Equity and Democracy, it would be the criminals who themselves pay for their crimes against us - humanity.

I also want to point out that Barack Obama, who campaigned on the slogan - 'Change we can believe in', also lied to us. As far as I can tell he changed nothing. None of the Bush initiatives were reversed in Obama's first 90 days in office as I expected. To me 'Change' of this nature would have meant at least that much. However, in his first two years in office, while had his party in complete control of Congress and therefore could have done anything he wanted to, he removed not a single soldier from Iraq or from any foreign country. He kept the NSA Spies spying on us. In fact, spying on Americans got worse under Obama's watch, not better. The national debt soared. He prosecuted none of the Bush co-conspirators who robbed all the banks and retirement plans of the entire world.

The only thing he changed was to legalize and complete the total monopoly that the Big Three Health Insurance companies hold over the American people by forcing all of us to buy their products, when it had been optional prior to Obama. At least he's finally talking about making Peace around the world in his last two years of his administration, but this is exactly when he should be doing the opposite and going after the Terrorists now running wild in the horrible political morass that Bush created in Iraq. Instead, he's made it worse by doing nothing to fight them.

Two ore three more quick examples:

11. ABOLISHING Genetically Modified Food with strict penalties for any farmer or corporation manufacturing such food. Ten years in prison and a fine of ten million dollars for breaking this law would just about do it. Over 75% of the food in America is now genetically modified. People are dying by the thousands, not from the GMO FOOD but from the pesticides that you use on the GMO Food. Also these chemicals are getting into the rivers and oceans of the

world and starting to kill all life on the planet. We have to reverse this dangerous trend pushed on the world by one of the largest corporations in the world - Monsanto - for the sole purpose of making more profits that get paid out to the chief criminals, the chief executive officers in the form of billions of dollars in their pockets.

ONE COMPANY - MONSANTO has the LEGAL RIGHT granted to it by CONGRESS to POISON US and the entire planet - they are even EXEMPTED FROM LEGAL LIABILITY BY LAW because they knew who to BRIBE.

This proposal would be an easy one to pass.

12. PERSONALLY - I'm sick and tired of all the MOTOR-CYCLISTS who are also granted the RIGHT BY LAW to POLLUTE MY ENVIRONMENT DAILY with their noisy machines. They can and should be MUFFLED like any other motor vehicle.

We should all vote on requiring all motorized vehicles to emit no more than 75 decibels of noise pollution. One of my pet peeves. I hate to have to listen to these cars and motorcycles that have intentionally installed mufflers that make them so noisy you can hear them coming a mile away, then for another mile as they go away. This is noise pollution at the cost of my peace and quiet and yours.

13. While we're on it, we should all vote to force all police, ambulance and fire response vehicles to stop using their sirens. These things are so noisy - the constant siren screams in our cities is driving us all mad. Lights should be enough. If they have to slow down a little bit on the way to an emergency to preserve the peace and quiet of our cities - I'm all for it. A few more people will die because of this no doubt, but as Spock said in Star Trek so eloquently - "The rights of the many outweigh the rights of the few."

14. We should all vote on limiting commercial messages on TV to no more than 4 minutes per hour of televised programming - regardless of the source. And, these four minutes must be shown at the beginning and end of every show, nothing in the middle. This way, we could all go to the john, or go to the refrigerator, make a phone call, use the remote control to mute them all - never being spammed by these advertisers unless we chose to do so and be able to watch TV in peace.

This next one is near and dear to my heart.

ENDING THE TYRANNY OF MONEY

15. Part A - No member of Congress shall accept payment or reward or benefit in any form from any person or business entity during their term of office. During a campaign, all contributions to any political party or individual election campaign shall be no more than $100. Any person or business entity found contributing any money or beneficial reward to a candidate or a Congressperson in excess of $100 in value is subject to a fine of no less than 10 X the value of the reward exceeding $100 in value and imprisonment of ten years in Federal prison. Any Congress person found to be accepting any financial rewards in excess of $100 per individual shall also be subject to a fine of 100 X the value in excess of $100 and a prison sentence of 20 years in Federal Prison.

Lobbying of any kind is hereby a FELONY punishable by ONE MILLION DOLLAR FINE per incident and 25 Years in Prison

VOTE NOW-

Part B
No Congressman or woman shall - upon leaving the Senate or House of Representatives shall take a job or accept money, or financial benefits of any kind over $100 in value from any corporation doing business in the United States where the sole purpose of that job is to influence Congress in any way or if the payment or reward is in payment for any legislation they have voted for or against while a member of Congress.

AS I RE-THINK this proposal - the way I've written it up in Part A - would make Part B unnecessary but I leave it to show you the evolution of my thinking on the matter. AND this is exactly how proposals will improve and evolve over time as the people become more and more familiar with this great tool for FREEDOM - the best one ever known.

This one I'm sure nearly every American would support because it would put an end to the corruption of money and the tyranny of Big Money. No longer would any politician promise one thing during the campaign and be influenced to do the opposite, as is the case in nearly every case, every subject, every policy, every law perpetrated in our political system today.

When money becomes only a small token of support, it would obvious take these types of contributions in the thousands and millions and therefore this influence becomes diluted and unimportant to the candidate and therefore can have no impact on the decisions of those who are forced to ask for money to run for their jobs.

It also stops the way that government is now just an arm of Big Business because all of them take jobs in the Industries that they were regulating and this is obviously a pay-off for favors done for that industry during their term in office. No one would believe for example that Dennis Hastert had acquired any knowledge of hedge fund investing during his career as a wrestling coach and later a House member. Yet this is how Wall Street rewarded him for favors he performed for the big banks and how he obviously got hold of over three million dollars that he later used to silence a person who was extorting him for illegal and immoral sex acts he accused him of, while a wrestling coach.

16. Making it MORE DIFFICULT FOR CRAZY PEOPLE OR CRIMINALS TO OWN OR CARRY A GUN

The politicians who take money from the GUN Manufacturers never propose anything that will kill their cash cow. But allow the guns to kill their fellow citizens, those they are sworn to protect and to serve - that's totally reasonable. The way is now clear for us to put this kind of GUN CONTROL on the National Ballot. Yes, we will still preserve the right in our Constitution to own guns - however, nowhere in the Constitution does it say that crazy people and criminals have the right to own guns. SO - we need a National Ballot Measure that says the following:

Anyone convicted of a felony in any of the states shall be disqualified to own guns and they will be punished if caught with any device that can kill another human being using an explosive charge with a fine of ONE MILLION DOLLARS and TWENTY YEARS in PRISON with no possibility of parole. Anyone reporting the offense to law enforcement shall receive 50% of the fine.

WE MIGHT FORMULATE THESE THREE either at the same time or over time.

In three simple strokes of the pen, we eliminate the THREE problem areas of the major violence in our society - **CRAZIES, CRIMINALS and GANGS**

SEE how SIMPLE DEMOCRACY can be when you take OUT the ability of the MONEY TO INFLUENCE the LAW-MAKING PROCESS?

WHY should we allow the NRA (National Rifle Association) a LOBBY of GUN MANUFACTURERS to RULE the LIVES and DIRECTLY CAUSE THE DEATHS of millions?

There is no reason for this. They can no longer hide behind the Fifth Amendment - which guarantees all Americans the right to 'BEAR ARMS' but it does not guarantee that same right to CRAZY LUNATICS and CRIMINALS. The founding fathers had no intention of allowing that to happen, but they could not foresee the days of LIMITLESS GREED AND CORRUPTION when CONGRESS would be so BLATANTLY BOUGHT AND SOLD.

Now we have three companion Ballot Measures that makes it EXTREMELY difficult if not IMPOSSIBLE for CRAZY MANIACS AND CRIMINALS to possess and/or carry a gun. It's states AS FOLLOWS:
Any person who has been designated as 'Mentally Incompetent' by a licensed psychologist, psychiatrist or mental health worker is disqualified to carry a weapon capable of killing another human being using an explosive charge. Penalty, upon conviction of carrying such a weapon while adjudgd to be 'Mentally Incompetent' shall be punishable by 20 years in prison and ONE MILLION DOLLAR FINE. . Anyone reporting the offense to law enforcement shall receive 50% of the fine.
Some people will always slip through the cracks no matter how tight you make them. But, this at least would be a huge deterrence. If a convicted felon shows up at a night club or bar and someone sees that he is carrying a gun, he is very likely to be turned in by a friend or onlooker who knows

that this could mean a cool PAY-DAY. That would keep guns off the streets in 90% of the cases.

Because we never get GOOD and EFFECTIVE laws on the books, we are forced to live with incidents like Sandy Hook, where one of the craziest of the crazies went in and slaughtered about thirty innocent children with an automatic machine gun. His mother had the gun in her house and this is how the crazy got the gun and the first thing he did was to kill his mother before driving over to the school.

We see things like this in America almost every day and they could mostly be prevented with good deterrent laws on the books. If this law that I crafted above was in effect, one of the mother's friends or relatives would have turned this kid in because he was certainly 'Mentally Incompetent' and there was a psychiatrist who had already determined that diagnosis on this crazy kid. So, someone would have pointed out to law enforcement that this kid was living in a home with these dangerous weapons. This is 'POSSESSION' and that whistle-blower would have saved thirty innocent lives and pocketed maybe as much as $500,000 for being observant and a good citizen.

It won't stop all the crazy murders of innocent people in this country, but this kind of law would certainly put a huge clamp on it.

Certainly - you would vote for it - I hope. I believe the vast majority of Americans, when they see all of the senseless violence on TV would vote to protect their own lives. In this case, it is NOT against the 5th Amendment - the one that guarantees us the right to own guns. It merely makes it more difficult for criminals and the crazies to own or be in possession of a gun where they can do so much harm to innocent people. The GUN LOBBY would not be able to BRIBE US ALL - and so this is the point of TRUE DEMOCRACY - the average voter decides the big issues of LIFE and DEATH - because we are the ones paying for them and we no longer have to rely on those THAT ARE BOUGHT AND PAID FOR by those who make money purveying DEATH.

17. A NATIONAL BALLOT MEASURE to make the DEATH PENALTY - MORE EFFECTIVE

OR - if a total Abolition is not acceptable to the people, we try something else.

OR - if this swipe at more justice is not acceptable to the majority we try something else.

    In my opinion, the Death Penalty should be reserved for crimes that are completely heinous in nature and are completely and utterly proven by scientific and/or forensic evidence of guilt of the person charged. NO ONE should be put to death in any civilized nation on the testimony of one or even two other individuals who many times have been easily mistaken - or worse are holding a grudge against the accused

OR may have made a deal with the Prosecutors for a lighter sentence in return for their testimony. This would not be JUSTICE as guaranteed to ALL under our Constitution.

We currently have known to have executed in this country over 2,000 people who were later proven INNOCENT of their crime. This should NEVER happen in a just and fair society. Too often the innocent are put to death on the basis of one witness - who could be lying to gain a lesser sentence. It happens all the time in the criminal justice system. This is corrupt and it is wrong.

Whenever, a person is found guilty by evidence such as DNA or fingerprints and a witness that can corroborate the scientific evidence - then in my opinion the death penalty may be applied.

Therefore - this Ballot Measure would OUTLAW any Death Penalty when there is no scientific evidence of guilt coupled with an eye-witness, placing the accused at the scene of the crime. Whenever we have both of these conditions in a trial, then and only then, may the death penalty be applied by the courts.

18. It shall be a felony for any police officer or any other member of government to shoot an American citizen in the back. If the effected citizen dies from the police shooting, this act shall be punishable by the immediate loss of their job, life in prison, without possibility of parole, and a fine of one million dollars that shall be used to compensate the family of the victims.

Another easy to understand law that would be supported by the vast majority of Americans is to make it a felony for any police officer, National Guard, US Army member, Air Force, Marines, etc. to shoot an American citizen in the back.

Today, there appears to be an understanding in the police departments of this country that it's allowable to shoot a citizen in the back. It used to be understood that it was defensible for a police officer to shoot someone if they felt their life was threatened. This would be self-defense. This law would not change that. It would only attempt to stop the insanity where someone who is hired to protect the public is out there slaughtering innocent citizens at will, most of the time due to the color of their skin. This is un-American and immoral and should be criminal. Obviously, whenever a police officer shoots someone in the back, that person would have to be running away from that police officer and is not posing any threat to the officer's life and should therefore be protected under the criminal justice system of a country that calls itself a country Of, By and For The People.

A tough law like this on the Federal level would over-ride any state laws that may be there to protect the police. This is true of all Federal Laws, especially those proposed and passed by the people.

AND I have another one I've just recently conceived from reading the news about all of the CORRUPTION IN US PRISONS and the SIZE of our PRISON POPULATION - the largest in the WORLD. This one would be so easy to implement and would be beneficial to society in so many ways and is therefore instructive of the process of how to use the Internet more effectively. It would not only save billions of dollars - it would save millions of lives - and it would help save our planet. But best of all - it would cost NOTHING to the taxpayers.

19: PRISON REFORM

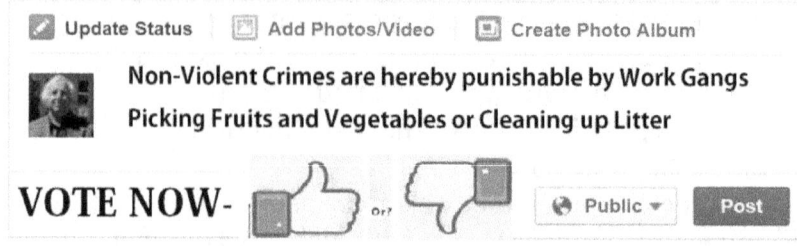

THIS ONE STROKE OF THE PEOPLE - would help SOLVE THREE MAJOR PROBLEMS in our society.

    A. The BILLIONS wasted on prisons
    B. The IMMIGRATION problem because we would have an instant labor force to pick our fruits and veggies - mostly ALL AMERICAN Citizens would be learning a viable job skill.
    C. The failing INFRA-STRUCTURE.

    AND the COST OF FOOD would go DOWN to boot.

It is herein provided that no more shall anyone convicted of a crime - other than a crime of violence - shall be sentenced to time in jail or prison - unless it is to house them and provide a place from which they will strike out into the community - six days a week - doing activities that improve the community such as:

    A. They will clean graffiti
    B. They will fix roads and bridges
    C. They will rescue lost animals
    D. They will fight forest fires and even building fires
    E. They will work to prevent forest fires.
    F. They will clean up parks, beaches, other public spaces.
    G. They will help feed the homeless by collecting food waste from restaurants.
    H. They will aid police and hospitals do their duties

I. They will pick up garbage and other refuse aid in recycling efforts.
J. They will help in reclaiming wetlands and other natural habitats.
K. They will perform such duties as a committee of concerned citizens determines to be in the public interest.

Would you vote YES or NO to such a proposal?

I believe most of my readers would easily comprehend the money-saving aspect of this proposal as well as the saving of human lives, the re-training, the gift of a real and useful job skill to a young criminal would be another major benefit, that would cost less than the cost of keeping the criminal is a 7 x 12 foot cell, guarding him all day and giving him three square meals a day without contributing to society in any way.

Today, we have millions of young people rotting in jails for years and their crime may have been smoking pot or selling pot to other bored teen-agers. Many in this predicament call their prison time 'Crime College' because they learn how to do more serious crime from the older inmates who have nothing better to do than proselytize the younger inmates. Many youth today actually desire a prison sentence so that they can attend this college of higher criminal activity and get a free education in how to damage society even further after they 'graduate'.

Isn't it a better idea to put them to work improving our streets and roads where they would not have any time to learn anything except skills that they can truly use to further improve society and become tax-paying law-abiding contributors to society for the rest of their lives?

I HAVE TOO MANY OF THEM for a full discussion here, so I am going to SHOW you a few more that I believe you can comprehend without any discussion of each.

YOU may come up with MANY MORE of your own solutions.

WHY do they USE OUR MONEY TO SPY ON US?
Because WE HAVE NO WAY TO STOP IT~

HOW would you vote on this one?

And this one?

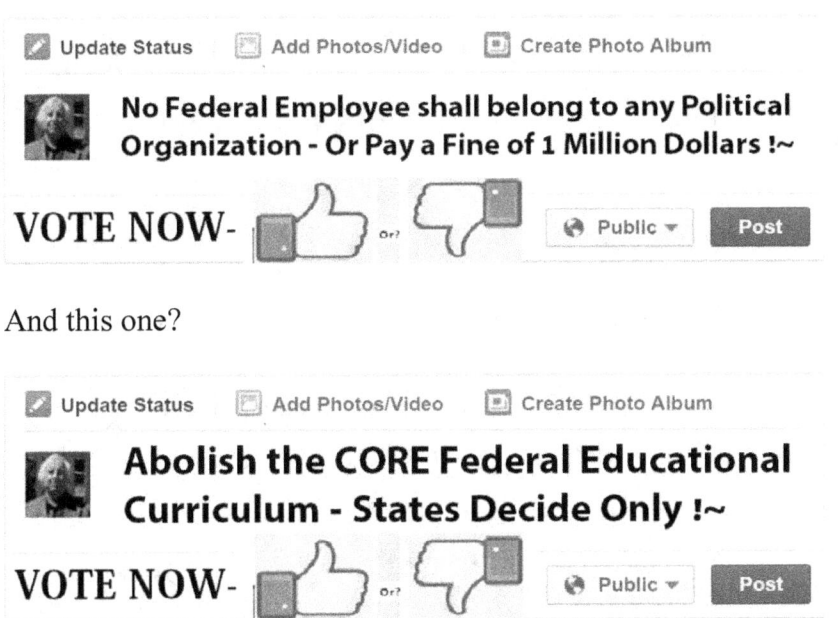

And this one?

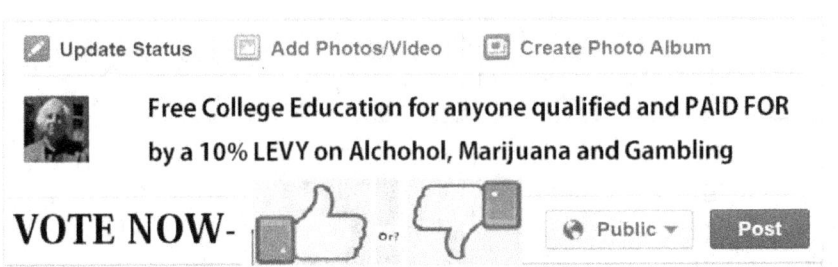

This CORE Educational System is too close to PROPAGANDA for my liking - how about you?

AND this is an easy one for most of us? Gambling alone in this country could pay for ALL college educations for everyone if taxed in this manner. Marijuana will soon be the second greatest source of revenue for this kind of thing.

Do you want to see drones hovering over your house and back yard?

   Obama wanted to give us SINGLE-PAYER but he says he was not powerful enough to over-ride the wishes of Congress in blocking Single-Payer Medical System very popular in most of Europe. Obama made this excuse even though he had his own party in the majority of CONGRESS, so that seems to be a very lame excuse to me.

   But, again, the reason that even Democrats don't vote the way a President of their own party wishes them to vote is that they are all on the take and taking money from the monopoly Health Care System who paid off our Congress to defeat Obama's wishes so that they could stay in business. Instead of being honest with the American people, Obama comes up with the worst alternative ever which gives the IRS more power to harass us and now they can do so if we don't buy health Insurance.

These are just a few examples of how - when left to the decision-making process of Too Few, we wind up with 'Too Late' and Too Expensive and truly costly solutions for the rest of us. If I - one person - can think up a better way to treat the criminal in society - imagine what we can all do when we put our heads together.

As John Lennon said - "Imagine - All the people living life in peace."

Of course - these would all be my own suggestions of National Ballot Proposals. You may have several ideas of your own far superior to mine and I would love to see them. I'm also sure that not all of these examples would pass with a majority. The anti-siren one for example, may not pass being that many people don't regard noise pollution as that serious and they may be more sympathetic to those in need of these emergency services than I am. But I would vote for such a rule of law because I'm hearing more and more of them and it's alarming to me - but perhaps most would not be as alarmed as myself. I don't require that all common sense issues pass this test, only that we have the chance to put them to the test of the will of the majority. This is the only form of government that makes sense to me and I hope you the reader.

What the Internet offers us, and especially through new found knowledge of The Force is a way to open up the political leadership of this country and later the world to thousands of candidates, instead of just the two mediocre ones that they put up for us by the two parties. Both parties in this country suffer from bribery related disease that has taken root over hundreds of years. If we don't get rid of the two-party system - the one true support of the Evil Empire, this country and the world may be doomed. In my humble opinion, the only way to get rid of them is to slowly vote them out, one by one, and the only way to do that is to open up the process to everyone who is not now, and never has been a member. Then, and only then, we will be able to elect someone to the highest office in the world who can actually govern with a sense of responsibility to the majority of us, and not just the lucky few.

This will only happen if we unleash the power of the collective Consciousness that I am hoping to unleash here and now. Get ready, America. It's coming.

All the rest that goes along with changing the world this systemically and thoroughly will stem from these examples that I have given where the American people come together, as one, seize the moment, analyze the momentum of each moment in history as they are given to us, and then making all the necessary course corrections.

By volunteering to be part of the Jedi Knight - Rebel Alliance, and if there are a critical mass sufficient to make it happen, all of this should become available to the American People soon so that we can then take the next great leap forward - The Constitution of the United Countries of Earth.

THE FIRST VOLUNTEER WAVE that I am requesting are PEOPLE who will get active in this 2016 Presidential Election by calling and emailing every one of them and telling them that you will not work for them or give them any money until and unless they support the idea of making the laws on the Internet - a Real Democracy in America. They will get it - at least one of them - if we have enough volunteers getting in their business every day.

Go to their rallies and call them out on it. Ask them if they would support Real Democracy over the Internet and listen to their equivocating patronizing answer. Then, point out to the crowd how they avoided the question. Do this over and over until you embarrass them all into either losing the race and preferably to someone who will join us.

Call the media and feed them stories about how you tried to get a candidate to endorse a Real Democracy in America and they refused. Use other tactics that you can conceive to force them to the truth.

Confer with others on the Internet for strategies and techniques that have been working. There is strength in numbers. This Knighthood is real and when we reach a certain number of us doing all of these things - we're going to change the world and save our country first and the rest of the planet will follow.

Get active before we're all radio-active.

Another class of early volunteers most sorely needed are computer programmers. If you have been following this experiment's orientation, we will need to do everything at the best possible speed and with the highest and best use of the technology at hand. I believe that the way to get started down the right path is to be able to offer the world a 'Facebook-Style' of website where we can all gather together to complete all of the steps required in this experiment. AND I have other ways to accomplish this as well

So, if you are a Facebook programmer, we need you. If you have Google experience, we need you now. If you have worked for Microsoft - we need you. If you work for Apple Computers or Dell, we need you. If you have phone app programming experience we need you. We need anyone urgently who has any ideas at all about how to save our planet using the highest best powers of the Internet.

~ ~ ~

CHAPTER FIVE

- Yelpocracy In Action

The first things we need you to do as a simple matter of educating the public on how this will work is to actually BEGIN USING IT in its first Beta Test Form - and the start of this process is to RATE THIS BOOK - at the website - please gives us a FIVE-STAR Rating plus any comments you'd like to add so that it can become personal to you and the rest of our readers.

If you give us a RATING and a Comment, we'll track you down and send you a really nice reward for all of your time and energy.

GO and visit several websites that relate to the ideas you just read about and GIVE IT A YELP RATING of 5 Stars - if you can.

The first one is this Book's Main Blog Site - http://www.Yelpocracy.com

But, at this stage in the Revolution, we need more from all or at least most of my readers. We need you to put this book first and foremost in every communication or most every communication you make with friends and contacts from now on.
We will NEVER get the support of the main-stream media because they favor the system we have now where they get a disproportionate share of the power over us. What we are recommending would make them just another VOTER like all the rest of us and their influence, as it exists today - vanishes, out the window, finito, gone.

Therefore, the only way we can accomplish all of what I am proposing here is if there is an historic grass-roots movement, the likes of which we have never seen before. Each and every one of us must learn to become a soldier in the defense of our freedom and liberty. We must begin to make the quest for a Real Democracy - in this case - the Yelpocracy - just as much a part of our daily lives as eating, breathing and sleeping.

We have to jump out of bed every day dedicated to the goal of bringing at least one other person into the fold. Some of you are capable of introducing this idea to dozens, others even hundreds and some of you can bring thousands to the fold. If you only can bring one other than yourself - you're doing your part. But, don't stop there. Learn the actions and words that brought in that one person and try it again on another. Then, another and another in succession until you have done everything you can in the time you have left on this planet.

This is the only way. Confucious said, "A journey of a thousand miles must begin with the first step." This is our first step collectively. We can get there. We must get there. We will get there. But, each of us must begin the journey toward Yelpocracy with the first steps. That very first step that we need from you is the step of RATING this book so that you become familiar with the process and you will see aspects of it that I have missed which will make you just as much of a founding father or mother as myself as you all begin to develop related concepts that make this task of using technology to rule the world fairly and justly - even easier.

# # #

The most important thing to remember about all of this is that YELPOCRACY STARTS RIGHT HERE - RIGHT NOW

WITH YOUR RATING OF THIS BOOK and THESE RELATED IDEAS AT OUR WEBSITE

www.Yelpocracy.com

PLEASE give this book a ~~

~~ 5-Star REVIEW ~~

AND if you can swing another 5 minutes another REVIEW on either Amazon or Audible, depending on how you received the book. Amazon allows any reader to review the book they purchased and when we reach a few hundred reviews by our volunteers, the book will become featured more in Amazon or Audible which makes it more known and on its way to best-seller status. So, after completing this book - please search it in Amazon or Audible once again and you will see the place where you can give it a 5-Star Review. If everyone reviews the book positively on Amazon - it will become well-known enough to kick start the process to reach the critical mass we need.

THIS IS THE BETA TEST - IF YOU CANNOT DO THESE THINGS to help save your country and heal the world - WHO WILL?

This is the first stage of HOW we can achieve so much and the reason that I have spelled it all out for you as an Experiment is that we now have all of these tools that we can use to get ourselves out of the worst pickle we've ever known in all of our 5 million year climb up out of the jungle and into cities and towns. By this point, you should be all fired up and ready to go. You are starting to feel the collective consciousness or if not - you soon will begin to feel it. Everyone catches on in their own schedule and in their own way. I know you are catching on and this is my greatest accomplishment to date and one for which I am the most grateful and humbled.

But, you need more direction. You may even have the Cosmic Connection in place already that I am trying to instill in every one of my volunteers. But, I also know that some do not have that yet and need just a little bit more encouragement - a few more details on how we can accomplish so much that we would be saving our planet and our own species from the ultimate and conclusive disaster.

Well, the answer to the biggest question in your mind is that we will use the power that we have just unleashed in 'You' and by that I mean the collective 'You' or all of you, the communicating and very powerful union of those of you who have volunteered to see this to the end. This is the 'Collective Consciousness' that we are testing here and it is now up to you to be part of it or to ignore it. Being part of it is now made easier by the invention of the Internet as I have said, but in this chapter I am going to give you how I can see the Collective Consciousness working in much greater detail and it is at this close-up magnification of this experiment that I hope to capture your participation once and for all time.

By placing our full faith and trust in the greatest tool for democracy to ever land here on planet Earth, we can and must start to turn our world from its present course to total annihilation and destruction of all life forms on this planet towards a final destination of total peace, harmony and global sustainable economy and husbanding of our resources in such a way that we guarantee all future generations of humans and all other life forms an always-improving environment upon which to live.

Please remember that this line of reasoning and the accompanying chores that we ask of you is based on the following question:

Why in the name of all that's holy - do we allow massive decisions effecting the entire planet - the fate of over 7 billion people and countless billions of plants and animals upon which we depend be made under the

mental capacity of only a few? Why do we rely on this antiquated method of governing ourselves when the tool for managing all of this for the benefit of all 7 billion of us is right before our eyes?

You will want the proof.

In a previous book - "Using Google Super-Vote to Create The Super States of America" I detail how an Internet democracy could work with almost no major alterations to the Internet and how we use it today. So, I won't take the time to detail the nuts and bolts of a Real Democracy using the Internet except to say that if we can use Google Super-Vote where American TV viewers vote on their favorite singing contestant on American Idol and other reality TV shows, then, it's obvious this same technology could be used to vote for our most important proposed solutions to our most pressing problems, not only for who should win a million dollars.

Another consideration that people have whenever I mention this is that we do not have a Democracy in America because it was technically set up as a Republic. Although this is true historically, the same history books prove to us that when people want a major change in their system of government, they merely band together and demand it. Eventually it gets done if they are persistent and have the courage of their convictions. Mohatma Gandhi, the man who single-handedly brought a form of democracy to India and freed his country from British tyranny said "First, they ignore you. Then, they laugh at you. Then, they fight you. Then, you win."

Let's all win - starting now.

In rating this book - you take Step One in the process of delivering True Freedom, Liberty and Justice for all to the American people. You become a Yelpocrat and you become a founding partner in the only sustainable future of this country and the world outside of our borders.

Upon completing this book - Please visit our website and make your rating count. http://www.yelpocracy.com

Start the Revolution

GIVE THIS BOOK YOUR RATING

The world will thank you.

- About the Author -

This book is the third installment of my Trilogy that is composed of three books which chronicle the greatest discovery in our long evolution as a species - the God Particle and the impact that it will have on our species over the long term.

They are: The God Particle Bible

The Force Equation

Which Inspired -

The ORIGIN of CREATION

Which Inspired -

The 4 States of Consciousness

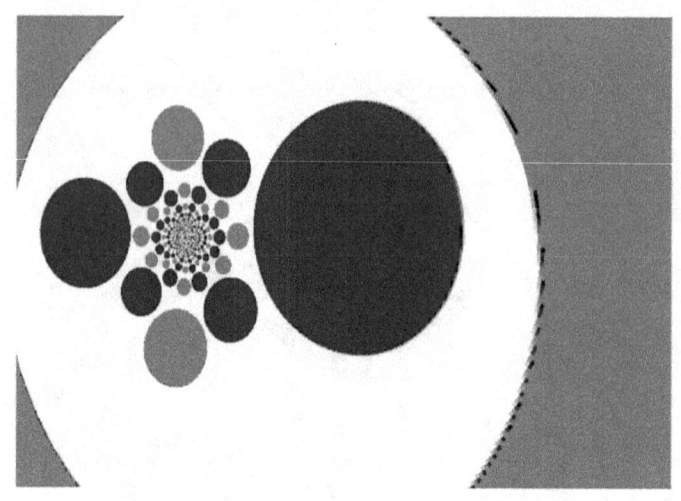

The Four States Of Consciousness

Michael Mathiesen

Which inspired this book - Yelpocracy

As a volunteer Yelpocrat in this great new and optimistic view of the future - I also ask that you read and review and recommend ALL FIVE of these books to become completely knowledgeable in the greatest scientific discovery of all time - or how the Human Race will evolve enough to meet our Creator's greatest expectations of us. Presently, as a species we are still children. When we fully mature enough to learn how to manage our world for the greatest benefit for the largest number of us at all times - we will have opened the door to God's greatest abundance and fulfillment of our spirit.

I JUST CREATED A FULL COURSE that I'm hoping will help this experiment in political revolution spread virally around the world.

TAKE THE COURSE IN CONSCIOUSNESS

~ ~ ~

I was born in Boston, Massachusetts to my two loving and caring mother and father Ann and Joseph Matarazzo - now in Heaven together - who gave me all the freedom I needed to think clearly - thank God. I was born precisely like all the rest of you, at the exact moment in time when God needed me here. Growing up in the seat of the greatest event in history - the American Revolution had its influence on me. My earliest memories as a young boy are almost always about my challenging authority.

Being raised in a Catholic household, I was subjected to huge amounts of dogma as it related to God. I knew, even in my formative years that they had it all wrong. What was written in the Bible seemed like a long series of strange fairy tales by people who had nothing better to do than to spin them a certain way that would make them all feel better. They were all about how someone hated someone else and killed them in revenge for an earlier crime, but that God loved them the most over and above all other people and that this was even justification for killing these other people - who did not have God's favor. Then, later someone, they claimed would be like 'God' or actually be the God of all time would come back to make it all right. The God of the Bible, it seemed to me had all the human flaws that make life so difficult in our society like jealousy, ego. etc. For example, one of the first stories of the Bible is how God tells a fellow called Abraham to kill his own son to show his respect and faith in God. Now, honestly, how petty and insecure would God be if he went around trying to get people to prove that they truly loved him?

So, I had my doubts and these doubts led to questions and these questions have led up to this Trilogy of Books of mine.

Even as a child I could sense this sort of thing from my religious indoctrination. Somehow I was able to separate fact from obvious fiction. And, as I grew older, I could see patterns in my life and moments in my life where God was definitely playing a vital role, always in the background and in the most subtle of ways, but he or she was definitely interested in how my life was going and doing his best to gently push me in the right direction. The signs were everywhere. And if I became more and more sensitive to these subtleties, I might eventually learn the entire game that God is playing.

In 1972, Richard (Tricky Dick) Nixon sent me a letter ordering me to join his troops in Viet Nam and help them destroy peasant villages, killing their livestock, their farmlands, even ordering me to slaughter innocent women and children in a lost cause of some kind that only a mental midget, more precisely a Paranoid Schizophrenic would find worthy. I made one of the most important decisions in my life. I seized the moment, saw the impact it would have on my life - I would certainly be killed or horribly wounded and I would have worse psychological scars of having participated in the murder of countless innocent men, animals, women and children.

I then made the necessary course correction and defied the President of the United States and the United States army and single-handedly without any aid of any firearms or a weapon of any kind other than the truth, I won the greatest victory of my life or anyone's life and learned at that moment, some of the truths that I am giving you here. I sent Tricky Dick Nixon a letter back telling him where he could stick his immoral, unjust, tyrannical and illegal orders. At that precise moment in my life, I was more powerful than the entire military machinery of the United States Army, Navy, Air Force and Marines. It was only recently, however, that I learned the importance of that power and how powerful any individual can become armed with only the 'Truth'.

~ ~ ~

.
Another of my greatest victories was that I cured myself of Type 2 Diabetes. I did it without the aid of this country's health care system. In fact, I did it despite them. I found from my own research that the cure for Diabetes is fasting. I tried it and within three months, the disease was fully in remission. I know the syndrome is still in my body, but as regards any disease, as long as there are no symptoms, and you are not forced to take any medication to control the symptoms, the disease is 'cured' or at least in remission, because there are no ill-effects of the disease unless one has a relapse. So, I still control my diet carefully and read the food labels of everything I eat to ensure that the syndrome does not come back. But, this also gave me another proof that The Force is strong in me. But, The Force, as I hope I have proven here, can be just as strong in any one of us if we only try. As Yoda said, 'Do or do not - there is no try.'

But by far and away, the most joyful of moments for me came on July 4th 2012, when scientists in charge of the biggest scientific experiment in history - The Large Hadron Collider in Switzerland, announced that they had indeed discovered the God Particle. I was off on the most rewarding adventure of my life, culminating in this book. There will be no more important books for me and my career.

On that auspicious and glorious day, I had landed in Maui Hawaii for a family vacation and I knew that this would be the most important few days of my life as I would finally be able to make sense of my entire life's events. As I bobbed up and down in the warm and azure waters of the beautiful and most blessed of all spots on this planet - I laid out the basis for my book - The God Particle Bible. All of that information was given to me as I bathed myself in the magical beauty of that precious island, smack-dab in the middle of the greatest ocean in the entire known universe. It was the greatest blessing of my life and one I will never forget saving, savoring and using that greatest energy to this day.

This experience was so smooth, so easy for me that my family didn't even know that I was working on the book that would change our lives. We had plenty of time for family outings. We went snorkeling and beach exploring every day of our two-week vacation. In the mornings before they woke up, I watched my dreams unfold onto the computer screen that thankfully we had taken along for the trip.

I now live in the second most beautiful part of this country - Santa Cruz, California with my beautiful wife Estrella and my two beautiful and loving daughters, Marissa and Kira in Santa Cruz, California, itself the most beautiful piece of the world I have ever known.

.

- GodParticleBible.com -

Makes a GREAT GIFT - Listen to the Book narrated by the author while you DRIVE to WORK or go about your daily chores.

As I mentioned above, much of what I have discovered here is through my prior discovery of the God Particle Field and how to put it to use in my daily life.

No understanding of The Force can be complete until one also learns as much as possible about the God Particle.

The God Particle has been discovered and we ALL need to learn more about it.

The discovery of the God Particle led me to realize that this was truly THE FORCE as described in the STAR WARS movies. So, I actually found the formula for how to use The Force and published. It's now a Mathematical certainty as to how much power you as the individual can unleash.

The Force Equation - Unleash The Power

And please CONTINUE the JOURNEY by reading my Science Fiction Novels based on all of the above -

Including:

Immortality - A True Story

NOT SCIENCE FICTION - but still FUTURISTIC

America 2.0 Inc. - Take Stock In America

Wherein I discuss in detail how we can and must incorporate the United States Government into a For-Profit corporation instead of the lose-as-much-money-as-you-can Non Profit that we must all suffer at the present time and further details on how to achieve a Real Democracy in America by voting on the Internet with many more examples of the use of The Force that you have learned in detail here.

I hope you have enjoyed this little notion of mine and that you will also enjoy other cutting edge creations that I have written. SOME OF YOU - are strong in The Force already. If you are, please tell me. I would so much like to know how many readers - use The Force already - what methods you may be using, things you wish to accomplish etc. Drop me an email at:

marketingcircles@Gmail.com

Also the author of 'One Million A.D. - The Story of Civilization 1,000,000 years from now.

And - 'Born Again - The 2nd Greatest Story Ever Told'

Also found on Amazon

Take the Course - It will blow your mind.

http://udemy.com/doomsdaywatch

For the Force to be with us in a sufficient quantity and quality - we have to reach critical mass.

Help spread THE FORCE

If you are already an AMAZON.Com AFFILIATE, please go to

Associates.Amazon.com

and search for this title. There you will find an AFFILIATE LINK for this book for example:

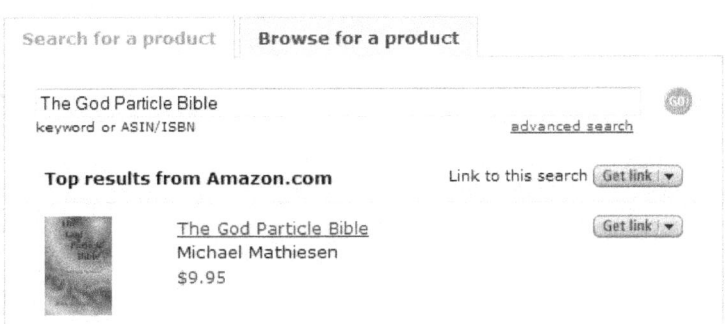

CLICK on **the YELLOW 'GET LINK' button** on the RIGHT and Amazon will give you YOUR AFFILIATE LINK for this book allowing you to then POST THIS LINK on your own websites and send in emails, etc. This pays you the Amazon standard rate of Commissions on any sales you make.

IF you are NOT an AMAZON affiliate as of yet

GO TO Associates.Amazon.com

And sign up and then search for your AFFIILATE Link for this book and market it to everyone you know. In this manner, we may reach Critical Mass sooner rather than later.

May The Force Be With us!

And please take the time to give this book and others in the Trilogy a nice REVIEW ON AMAZON or AUDIBLE.COM. This is part of how we save the world.

www.ingramcontent.com/pod-product-compliance
Lightning Source LLC
Chambersburg PA
CBHW071822200526
45169CB00018B/636